PROPANE CONVERSION OF CARS, TRUCKS & RVS

Other TAB books by the author:

No. 2084 *How To Make Your Own Alcohol Fuels—2nd Edition*

PROPANE CONVERSION OF CARS, TRUCKS & RVS

BY LARRY W. CARLEY

TAB BOOKS Inc.
BLUE RIDGE SUMMIT, PA. 17214

FIRST EDITION

FIRST PRINTING

Printed in the United States of America

Library of Congress Cataloging in Publication Data

Carley, Larry W.
Propane conversion of cars, trucks & RVs.

Includes index.
1. Automobiles—Motors (Compressed gas)
I. Title.
TL229.G33C37 629.2′504 82-5684
ISBN 0-8306-3103-8 AACR2
ISBN 0-8306-2103-2 (pbk.)

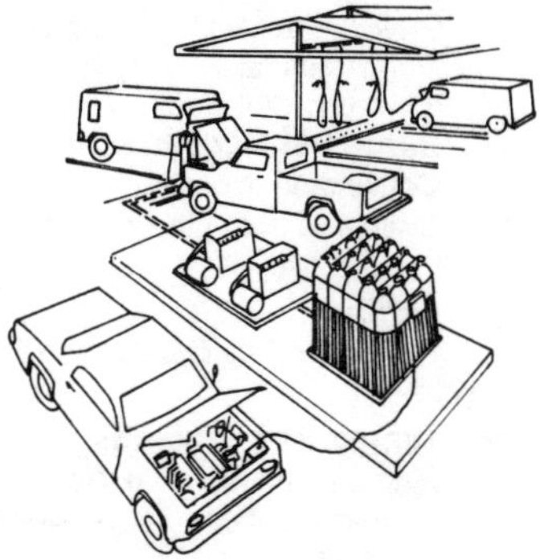

Contents

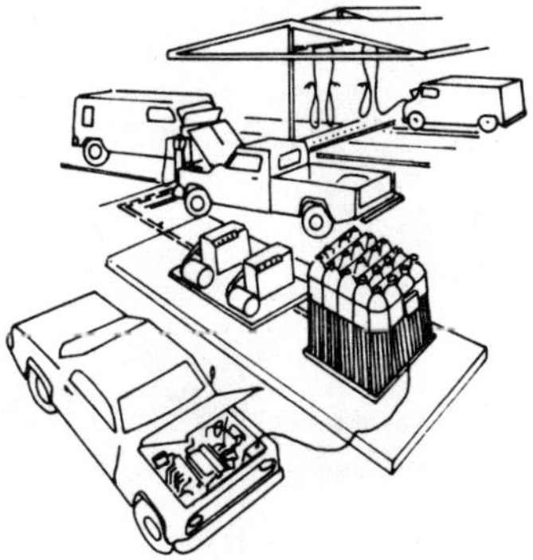

Acknowledgments

The author wishes to gratefully acknowledge the help provided by the following in preparing this book:

Dual Fuel Systems Inc.
Impco Carburetion
J.J. Parkinson Co., Inc.
NAMOCO
National LP-Gas Association
Petrolane LP-Gas Division
The Dancey Co., Inc.

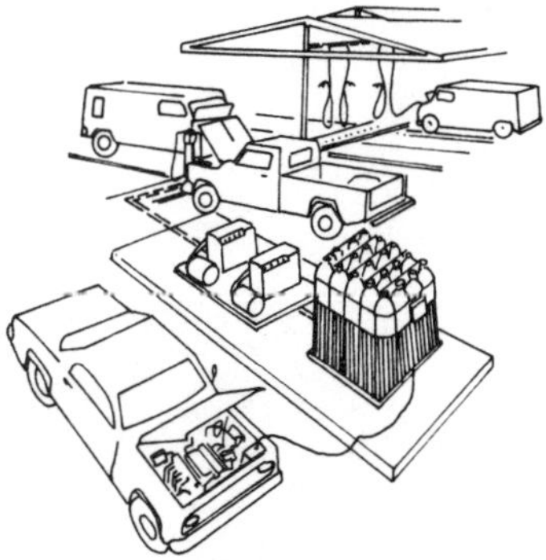

Introduction

This book explains the merits of propane and natural gas as alternative motor fuels and describes the basic steps involved in converting gasoline-powered automobiles, trucks, and recreational vehicles to run on these fuels.

Although all the essential elements are examined, a propane or natural gas conversion is *not* something a novice mechanic should attempt without supervision or guidance. Propane and natural gas are highly flammable, pressurized gases. One mistake while doing a conversion, resulting from ignorance, carelessness, or whatever, can transform an innocent-looking vehicle into a potentially lethal rolling bomb. Because of this, many propane- and natural-gas component manufacturers will not sell conversion equipment direct to the general public. They prefer to sell through distributors who install the equipment for their customers.

Some of these conversion centers also offer training courses for experienced mechanics who want to learn how to do their own conversions. If you are interested in converting your vehicle, you would be well advised to attend one of these classes.

Your state LP-Gas Association can help you find a local distributor for parts, service, and training. State associations also maintain lists of propane outlets in your area. A directory of nearly 25,000 propane dealers nationwide is available from the National LP-Gas Association in Oakbrook, Illinois.

When properly installed, a propane fuel system or a natural-gas fuel system provides a safe, dependable alternative to gasoline—and at substantial savings. The payback period for the initial conversion depends on the number of miles driven; in most cases it is a year or less. It is definitely an alternative worth considering.

Chapter 1

What Are the Alternatives?

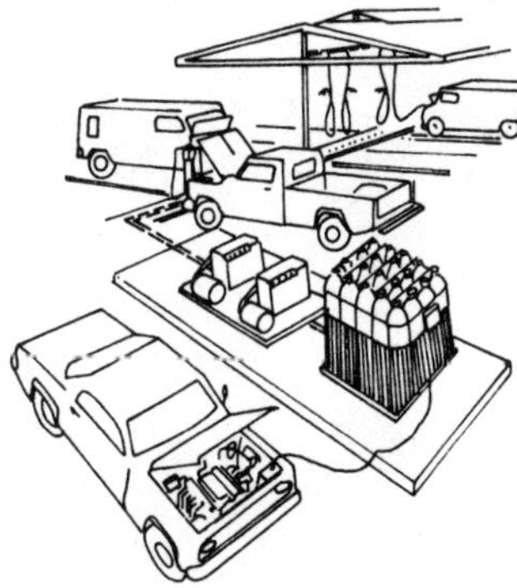

This book is about two fuel alternatives: propane (also called LP-gas or LPG) and natural gas. In recent years, these two fuels have become increasingly attractive to owners of car and truck fleets, recreational vehicle owners, farmers, and motorists thanks to the soaring prices of gasoline and diesel fuel. We all know what the Organization of Petroleum Exporting Countries (OPEC) has done to the price of crude oil and the disasterous impact those price hikes have had on our economy, our transportation system and our standard of living. Our only choice is either to pay the price and make do with less or to switch to alternative forms of energy.

No one can accurately predict how much we'll be paying for gasoline and diesel fuel next year or the year after. Neither can anyone accurately predict whether we'll have abundant supplies or another shortage (Fig. 1-1). The politically unstable Middle East, the wheelings and dealings of OPEC and our own Big Oil companies, and the uncertain nature of politics makes long-range planning extremely difficult. The only certainty seems to be a continuing trend towards higher prices for conventional fuels.

Propane and natural gas are fuels that can be practical and economical alternatives for many of us today. Both have proven to be good motor fuels, and both are considerably cheaper than gasoline or diesel fuel. In the summer of 1979, for example, propane was selling for about 30 cents a gallon less than gasoline. That year the number of vehicles being converted to propane doubled from

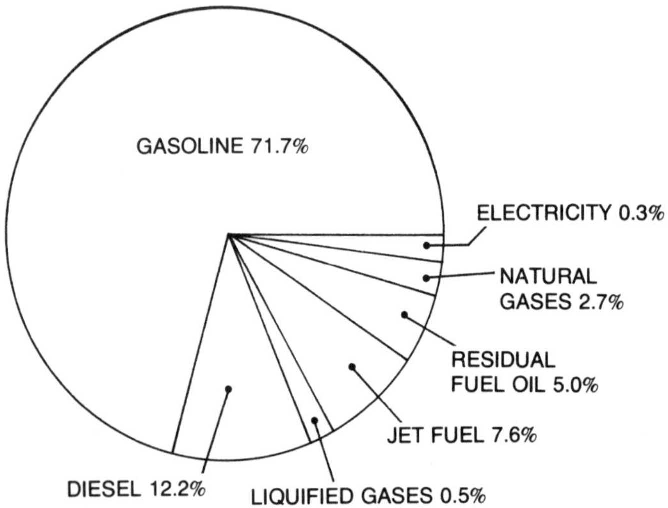

Fig. 1-1. Fuel consumed by various forms of transportation in the United States.

35,500 a year to 70,000 a year (Figs. 1-2 and 1-3). By the fall of 1981, gasoline and diesel fuel had risen to approximately $1.30 a gallon. Yet propane was selling for 70 cents to 80 cents a gallon in most parts of the country. The difference of 50 cents to 60 cents a gallon in propane's favor was enough to stimulate a whopping increase in the number of propane conversions.

According to industry estimates, 350,000 vehicles were converted to propane in 1981 (10 times as many as in 1978). Those who made the switch enjoyed some real savings in their fuel bills. In most cases, the conversion paid for itself in a year or less. Generally the price of propane is about two-thirds that of gasoline. The actual price the consumer pays in a particular area depends on the availability of the fuel, market competition and taxation by federal, state and city governments.

The main reason for the lower price of propane is that 70 percent of the propane produced comes from natural gas wells (the rest is a by-product of oil refining). Because the wellhead price of natural gas is controlled by the federal government, the price cannot rise as quickly as that of world crude oil. What's more, about 90 percent of the propane used in this country is produced right here in the United States, and that means the price falls under federal regulations. Taking such factors into account, it would be fairly safe to say that propane should continue to enjoy a significant price advantage over traditional fuels for a number of years to come.

How long will that advantage last? Those in the industry say there's nothing in the immediate future to change the situation. The Reagan administration has proposed completely deregulating natural gas prices by 1985. As with deregulation of the oil industry, the windfall profits would be used to explore for new wells. There were an estimated 165,000 natural gas wells in the United States in 1981. These wells supplied about one-fourth of the nation's energy

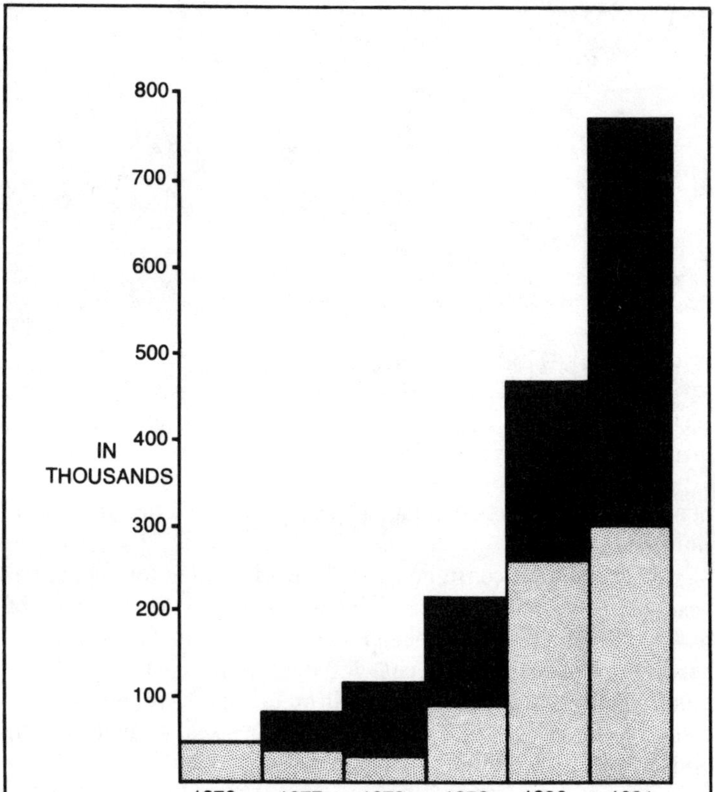

Fig. 1-2. LP carburetor sales for cars, trucks, and buses from 1976 through 1981.

Fig. 1-3. Shortages of gasoline in the 1970s and soaring prices in the 1980s are leading more and more people to seek out alternative fuels.

needs (only a small percentage of which was used for transportation).

Each year we consume about 20 trillion cubic feet of natural gas, and for years our consumption has outpaced the discovery of new wells. The result has been a steady decline in our natural gas reserves. Nevertheless, recent discoveries in the Rocky Mountains, off the coast of Louisiana, and in other areas have reversed the trend of dwindling reserves. Those discoveries, combined with public support of continued price controls on natural gas, seem to indicate complete deregulation has little chance in the years ahead.

Outside the United States, there is a world glut of natural gas and propane. In many oil-producing nations, natural gas is simply burned off at oil wellheads rather than recovered because the pipelines and facilities for storing and exporting the gas do not exist. These countries realize that they are burning money. Saudi Arabia has invested millions of dollars in recent years to build liquified natural gas (LNG) storage facilities and export terminals. By 1983, the Saudies will have an export capacity of nearly 20 million metric tons of propane. That is roughly equal to 80 percent of all the propane sold in the United States in 1978. Mexico, Canada and Venezuela are also boosting their propane output. By 1990, 5 billion to 6 billion gallons of propane a year might be available from these countries.

If the United States decides to supplement its existing natural gas supplies with increasing foreign imports and if plans to build a new pipeline to tap the gas reserves in Alaska don't fall through, the natural gas and propane supply situation should remain quite good through the rest of this century.

SHORT-TERM AND LONG-TERM ALTERNATIVES

Some of the alternatives to high gasoline prices that are available today include diesel fuel, alcohol fuels, hydrogen, synfuels, and electricity.

Diesel Fuel

Although you can't make a gasoline engine run on diesel fuel, you can convert to diesel power by replacing the gasoline engine with a diesel engine or by trading for a new vehicle with a diesel engine already in it. Although diesels have traditionally been used in long-haul, heavy-duty trucks, they have become popular in recent years because of their high fuel mileage. Diesels usually jet about 25 to 40 percent better mileage than a gasoline engine of similar size. But diesels have their drawbacks. They're noisy, they emit more exhaust pollutants, and they often smell awful. Some of the new automotive diesels, notably General Motor's 350 CID diesel, have suffered from numerous mechanical problems. And when something breaks on a diesel, it's practically always more expensive to fix than a gasoline engine. This is especially true of the diesel injectors and injection pump.

Diesels in automobiles and light trucks also typically suffer from a lack of power. Some engineers have commented that the same mileage improvements could be achieved by simply installing an underpowered gasoline engine. This approach would be less expensive and cause fewer headaches.

Diesels also require more maintenance, in some respects, than gasoline engines. Although there are no spark plugs to replace, oil and oil filter changes must be done about twice as often as a gasoline engine because of the diesel's high compression. This causes sooty combustion by-products to be blown past the piston rings and into the crankcase. The oil quickly becomes contaminated and it should be replaced every 3000 miles.

Switching to diesel power also involves a price penalty. Replacing a perfectly good gasoline engine is an expensive way to save a few bucks off the fuel bill. The same applies to the purchase of a new diesel-powered car or truck. The diesel option usually adds $1000 to $1500 to the sticker price, and unless that vehicle is going to be driven enough miles to offset the initial price difference through fuel savings, the diesel isn't worth the investment.

A final item to consider is that diesel fuel, like gasoline, is refined from crude oil. As we all know, crude oil prices are subject to the whims of OPEC. We can undoubtedly expect periodic price hikes and occasional shortages. What's more, refiners can't produce as much diesel fuel from a barrel of oil as they can gasoline. You get less fuel per barrel of oil when you switch to diesel.

Propane is probably a better alternative than diesel for many applications because:

- An existing gasoline engine can be easily modified to burn propane.
- Propane is a clean-burning fuel that emits few pollutants.
- Oil change intervals and spark plug life are greatly extended on propane.
- Propane fuel systems are typically very reliable and require little maintenance.
- The cost of a propane conversion pays for itself more quickly than that of a diesel conversion.
- Propane is less expensive than either gasoline or diesel fuel, and it gives roughly the same mileage as gasoline. There is no loss of power or performance with propane.

Alcohol Fuels

This is a subject that's received a great deal of publicity recently. There are basically three alcohol fuels that are considered as practical alternatives to gasoline. *Gasohol* is a blend of 90 percent gasoline and 10 percent alcohol. Straight *ethanol alcohol* is made from grain. Straight *methanol alcohol* which is made from wood, natural gas, and coal.

Farmers are interested in ethanol alcohol for two reasons. It's another market for their crops and it's something they can make for themselves (see TAB book No. 2084, *How To Make Your Own Alcohol Fuels-2nd Edition*. Brazil has adopted a national alcohol fuels program and hopes to have most of its vehicles converted in the near future. It appeared as though we would have a sizable ethanol alcohol industry in our own country under the Carter administration, but President Reagan pulled back the federal support of the synfuels program so that free enterprise—make that Big Oil interests—could determine our energy future.

Methanol alcohol, like ethanol, has been around for quite some time. It's been used in Indy 500 race cars for years. Recently, however, methanol has been receiving support from the oil companies as a possible alternative because it can be made from natural gas and coal. Because the major oil companies own more coal reserves than they do oil wells, development of a methanol alcohol industry seems like a distinct possibility.

Unfortunately, the only alcohol fuel that's available in quantity to the general public is gasohol. And burning gasohol is like burning gasoline. The price is about the same (a few cents higher in some states, a few cents less in others), performance is about the same, and mileage is about the same. The only real benefit of using gasohol is that it has a higher octane rating that helps to reduce pinging in many newer cars, and it could extend our gasoline supplies by approximately 10 percent.

As for ethanol and methanol, you can't buy either at the corner service station. Unless you have a still to make your own ethanol or a source of supply, you can forget about alcohol as an alternative for the time being. The economics aren't that attractive yet unless you can make the fuel yourself (and that's impractical for most motorists). Commercial alcohol was selling for about $1 per gallon in 1981. Though less expensive than gasoline or diesel fuel on a per gallon basis, an engine typically requires 20 to 30 percent more ethanol, or 50 percent more methanol, to run the same number of miles. The higher fuel consumption more than offsets the less expensive per gallon price.

Until alcohol becomes widely available at a competitive price, its use will be limited to self-sufficient farms and small fleets. Propane's advantage here is that there are 25,000 dealers nationwide and the price is far more competitive.

Hydrogen

Like propane and natural gas, hydrogen is a good gaseous fuel. It is clean burning, efficient, and it has been successfully used as a motor fuel in a number of experimental applications. Hydrogen's big drawback is cost. Even though hydrogen can be made by splitting water molecules in two with electricity, the cost of the electricity makes the hydrogen too expensive to be a practical alternative. Much work is being done to find an economical method of making hydrogen out of seawater using solar cells. The cost of solar-produced electricity is coming down, but it has a long way to go before it is competitive with existing utility rates. Until someone makes a big breakthrough in lowering costs, hydrogen will remain a promising possibility.

Using hydrogen as a motor fuel also has certain limitations. Like natural gas, it must be highly compressed for storage. Compressing the gas is no problem (although it does cost money to run a compressor) and storing it in heavy steel cylinders is no problem. The problem is cramming enough of the heavy steel cylinders—that are about the same size as large acetylene welding tanks—into a vehicle to give it a practical driving range. A small passenger car fitted with two hydrogen tanks would have a range of about 60 miles. That is sufficient for short commuter trips or local fleet use, but not enough for long distance driving.

If the cost of hydrogen comes down at some future point, it could, like natural gas, be used as a viable alternative by certain fleets. Research is also being done on alternate methods of storing hydrogen (including metal compounds that soak up hydrogen like a sponge). Unfortunately, these systems have proven to be as heavy and bulky as the traditional heavy steel tanks. Again, a breakthrough here could make hydrogen more attractive as an alternative motor fuel.

Propane, by comparison, does not present the storage problems of hydrogen or natural gas. Unlike the other two gases, propane condenses into a liquid when it is compressed. This allows more fuel to fit into a smaller tank. Because the pressures are much lower (175 psi for propane versus 2400 psi for natural gas or hydrogen), the fuel tank can be much lighter in construction. Thus a propane fuel tank of comparable bulk can provide a far greater driving range. For instance, a fairly large propane tank such as a 20-inch-by-64-inch, 80-gallon tank mounted in the bed of a pickup truck would provide a driving range of over 1000 miles on a single fill up. Empty weight of the tank would be about 240 pounds. In contrast, two 60-inch-long hydrogen tanks would weigh about 280 pounds, and would only provide about 60 miles worth of driving on a single fill up.

Synfuels

The United States has vast deposits of oil shale and coal that experts say can be converted into synthetic gasoline or diesel fuel to meet our energy needs. Consider *oil shale* for example. It is a type of rock that contains a substance known as *kerogen*. When crushed and heated, kerogen can be transformed into crude oil. The oil can then be refined, using conventional technology, into gasoline or diesel fuel.

In the Piceanoe Basin, in Rio Blanco County, Colorado, there is an estimated 480 *billion* barrels of recoverable oil locked up in oil shale deposits. That's more oil than in Saudi Arabia, Kuwait and existing United States oil wells combined! The only stumbling block to tapping this vast reserve of energy is that it takes a ton of rock to make just one barrel of oil! What's more, that ton of rock must be crushed and heated to 900°F to free the oil.

Engineers have developed two approaches to recover the oil and both of them are very expensive. One way is to simply mine the oil shale from the earth, crush it, and process it in giant kettles in special processing plants. A commercial plant, engineers estimate, would have to process 50,000 tons of rock a day to make money. The process also has its drawbacks. There are problems with what to do with the tons of waste rock that are left when the oil has been removed. There are also problems with obtaining the huge quantities of water that are required to make the process work. Water is already a scarce resource in the Western states. Having to share what little water there is with a shale oil plant would surely arise the anger of many Westerners.

An alternative way to recover the oil is to crush the rock underground and cook it in place in man-made caverns. This eliminates the problem of what to do with the waste rock and avoids the use of water. Experimental mines have fallen short of expectations.

The oil industry estimates that once the bugs are worked out of the process, oil can be recovered for around $36 a barrel. That is about the same as what we are paying OPEC for Arabian crude. The government is less optimistic. They say price would probably be $42 to $58 a barrel or more. And that is not counting the costs of environmental impact and cleanup measures. By comparison, the Arabs can pump a barrel of crude oil from the ground for about 40 cents. Synthetic gasoline or diesel fuel from oil shale, therefore, would be extremely expensive in relation to fuel costs today.

Coal is another prospect for synfuel production. Like oil shale, the United States has vast reserves of coal. If converted to oil, proven reserves in 1981 would total around 970 *billion* barrels of oil. That is more than double the entire world's proven oil reserves.

Making coal into gasoline involves a process called *liquifaction*. There are nine different coal liquifaction processes that could be used, and all achieve the same end through somewhat different means. The choice depends on the type of coal used.

Synthetic gasoline from coal has been around for quite some time. The oil-short Germans pioneered the process during World War II to help fuel their military machines. In South Africa, synthetic gasoline from coal has been a reality for more than 20 years. We even have a number of operating coal synfuel plants in this country; our largest one is in Catlettsburg, Kentucky.

Coal has its drawbacks. Extracting coal usually involves strip mining, and that is a method despised by environmentalists because of the severe damage to the landscape. There is also the threat of water pollution from mining and plant operations. These add to the costs of an already expensive process. But like oil shale, it is one of the few options open to us for extending our supply of petroleum fuels.

Remember that synfuels are expensive substitutes. The decision to use or not to use synfuels is not up to the individual. They will become a part of our energy supply when economics and politics justify the cost. The federal government has already agreed to partially subsidize the synfuels industry by contracting to buy a large percentage of the synfuel from the Colorado experimental projects.

The alternative to high-priced synfuels and conventional fuels is propane. Propane comes primarily from natural gas wells. Because the price is controlled, it should remain less expensive than either conventional fuels or synfuels. Equally important is that neither propane or natural gas require complicated or environmentally dangerous technologies to produce.

Electricity

Another much-talked-about alternative to petroleum fuels is electric power. Electric vehicles are being used successfully by some fleets for short-trip, urban driving. Their operating economics are fairly competitive with those of conventional gasoline-powered vehicles. They cost about the same when you figure in initial purchase price, battery maintenance, battery replacement costs, and recharging.

Electric-powered vehicles will undoubtedly play an important role in the future, but today there are only a handful of companies in the world that manufacture electric vehicles. General Motors has announced its intention to produce an electric commuter car sometime in 1985 or 1986. In the meantime, engineers are working hard to improve the batteries and motors that will power the "volts-wagon."

State-of-the-art electrics still leave a lot to be desired (although the art is rapidly improving). Conversion costs are quite high and battery performance is anything but spectacular. Most electrics are lucky to get 70 miles on a charge—and that's at a snail's pace. Long recharge times can also be an inconvenience.

Until there is a significant breakthrough in battery design, propane is a far better choice for beating the high costs of gasoline. Propane conversion costs are far less, the payback period is fairly quick, and range, tank refilling, and performance are not problems.

THE PROPANE ALTERNATIVE

In spite of its many advantages over conventional fuels, propane is *not* everyone's ultimate answer to high gasoline prices. In 1980, there were some 12 billion gallons of propane sold in the United States. Of this, only about 1 billion gallons, or slightly more than 8 percent, was used as motor fuel. The rest was used for such purposes as space heating, grain drying, cooking, and industrial processes.

Because the United States has approximately 150 million vehicles on the road, there simply wouldn't be enough propane to go around if every car, bus, and truck were converted from gasoline or diesel fuel. Experts say that, if that were to happen, we'd burn up all the 12 billion gallons of propane in less than three months! Then we'd either have to import propane like crazy or come to a screeching halt.

Just because there isn't enough propane for everyone doesn't mean there isn't plenty for many of us. Based on current supplies, there is more than enough propane to meet the demands of every fleet in the country. That includes every taxicab, police car, rescue vehicle, fire truck, postal delivery truck, municipal and state government car pool vehicle, city bus, light delivery truck, and commercial vehicle on the road. There'd still be enough fuel for recreational vehicles, farm pickups, and many private automobiles.

Europeans have been converting vehicles to propane by the thousands. And why shouldn't they? For years, Europeans have been paying two to three times as much for gasoline as we have. Our gas prices seem incredibly low to most Europeans.

In Holland, 50 percent of all commercial vehicles and 8 percent of all their passenger cars have been converted to propane. In West Germany and England, there are self-serve LP-gas pumps at many service stations. There's no reason we can't do the same in this country. Although we currently have less than a half a million vehicles running on propane, we could and should be able to increase that number tenfold.

Although automobile manufacturers have not yet decided to offer a factory propane option on new cars, both Ford and International Harvester offer propane options on certain medium-duty trucks (Figs. 1-4 and 1-5). In 1981, such factory installations accounted for about 20 percent of the propane-powered vehicles on the road. The rest were aftermarket conversions.

In its promotional literature, Ford says the advantages of being able to buy a new truck from the factory already modified to run on propane are as follows: reduced conversion costs because the original carburetor and fuel system does not have to be replaced; better overall performance and fuel economy because the engine is engineered specifically for propane; and factory warranty protection.

Ford says propane can be an excellent alternative for truck operators who are looking for ways to reduce their operating costs. Propane is ideal for high-mileage applications as well as high-idle-time, stop-and-go driving. Some of the best candidates for propane include metropolitan delivery vehicles, garbage trucks, and school buses. Ford adds that their LPG-fuel system offers lower fuel costs, longer engine life, low exhaust emissions, reduced engine maintenance, and good performance.

Propane-fueled forklifts are often chosen over electric-powered forklifts because they're better suited to extended periods of heavy use. A propane forklift can be quickly refueled, but an

Fig. 1-4. Propane-powered trucks are rolling off the Ford assembly lines. Ford says the LP-gas system is available on all its medium-duty trucks.

Fig. 1-5. International Harvester's Cargostar had one of the company's first factory-installed propane fuel systems.

electric must be recharged overnight. Many of these forklifts are operated inside warehouses and factories, and propane's clean-burning qualities avoid exhaust emission problems that would result from burning gasoline.

FLEET EXPERIENCE

Hundreds of fleets have made the switch to propane. The Los Angeles *Times*, for example, started converting some of their delivery trucks during the early 1970s. By 1981, the *Times* was operating over 200 delivery trucks on propane. They converted to propane to save money and to insure an adequate supply of fuel. During the late 1970s gas crunch, California was plagued by numerous gasoline shortages. Propane, on the other hand, was in abundant supply. As one of their fleet supervisors says, "Availability of fuel is a key consideration in the publishing business. There isn't a big demand for three-day-old newspapers."

Interstate Brands, a company headquartered in Kansas City, delivers fresh bakery goods. The company has converted about 1000 of its GMC and Ford step vans to propane. Like the Los Angeles *Times*, their motivation for using propane is its cost advantage and availability. According to the company's records, operating costs on propane have been about 8.6 cents a mile compared to 12.2 cents per mile for gasoline. They have also found that propane-powered engines run 180,000 to 200,000 miles before requiring an overhaul. Their gasoline-powered vehicles operated for 80,000 to 90,000 miles before requiring overhaul. Oil drain intervals on their propane vans have been increased to 50,000 miles and spark plugs last at least 40,000 miles.

The Sheriff's Department of Levy County, Florida was that state's first law enforcement agency to operate an entire fleet of propane-powered police cruisers. The police department was faced with the choice of cutting its operating budget or cutting back on services. They chose to switch rather than fight. When the changeover was made, the department was able to buy propane for 54 cents a gallon compared to $1.21 for gasoline. The tremendous fuel savings paid for the conversion costs in less than three months.

In Petoskey, Michigan, Parker Motor Freight operates on propane a fleet of about 90 heavy-duty trucks. They made the decision to switch in 1979. At first they were somewhat leery because of the extremely cold winters in northern Michigan. Since converting, however, the company has had no trouble starting its big trucks despite temperatures that have fallen as low as –29°F. One company official said that, unless the battery went dead, he could leave a truck sitting outside in the cold for a month or more and be able to start it right up with no trouble. The company also found that their trucks got much better fuel mileage on propane. Some of their big 477 Fords were only getting about 2 to 2½ miles per gallon on gasoline. When switched to propane, the same engines picked up one-half to 1 more mpg. The fuel savings paid for the costs of conversion in six month's time. The drivers also found that the trucks ran better and had more pulling power on long grades.

In Greenwood, Arkansas, the local school district converted 19 of its 21 school buses to propane in 1976. They began saving from $200 to $400 in fuel costs per week. Total fuel savings from 1976 through 1981 were estimated to be between $40,000 and $70,000. Other benefits reaped from the conversions were reduced maintenance costs, 40,000-mile tune-up intervals, fewer muffler replacements, and elimination of fuel pilferage by employees. The District's school buses use a dual fuel system that allows them to use either propane or gasoline with the flick of a switch.

The City of San Antonio, Texas has converted several hundred of its vehicles to propane. These include garbage trucks, light-duty pickup trucks, dog catcher trucks, airport security patrol cars, administrative vehicles and several street sweepers. City officials say the conversions have saved local taxpayers millions in fuel and maintenance costs. Examples like these and hundreds of others just like them are proof of the dramatic savings and benefits that can result from converting to propane fuel. See Fig. 1-6.

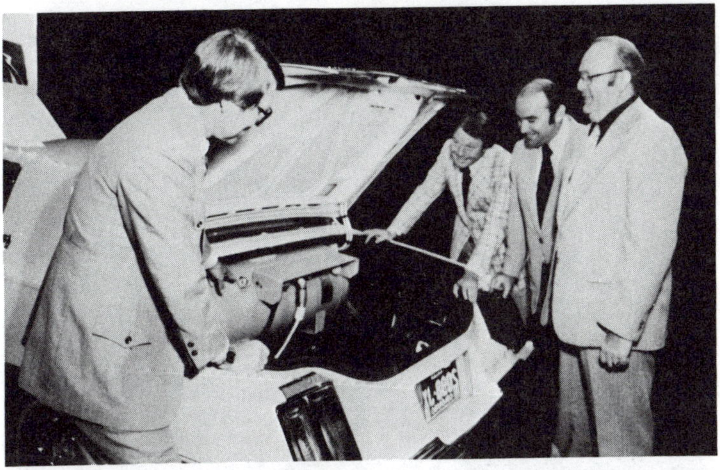

Fig. 1-6. McCauley Oldsmobile/Honda of Colorado Springs is cashing in on the propane business by offering its customers duel-fuel conversions on new cars.

THE NATURAL GAS ALTERNATIVE

Natural gas, like propane, can also be a very economical alternative to paying high fuel prices. The reason it hasn't achieved the popularity of propane is because it must be compressed to roughly 2400 psi. This is necessary in order to squeeze enough fuel into a fuel tank to give a vehicle a usable driving range. Compressing the gas requires an expensive compressor, and the fuel tanks must be made of heavy gauge steel to withstand the pressure.

Natural gas can be liquified like propane, but only if it is cooled to minus 259°F. This requires special refrigeration equipment and compressors. Once liquified, the LNG can be transported or used to power a vehicle provided it is held inside a heavily insulated fuel tank. These tanks are also quite heavy so the amount of fuel that can be carried by a vehicle is limited.

A rough rule-of-thumb engineers use when estimating the driving range of a natural gas conversion is that a compressed natural gas (CNG) vehicle will have about one-sixth the range of a similar vehicle powered by gasoline, and a liquified natural gas (LNG) conversion will have about half the range of its gasoline counterpart.

In 1969, the General Services Administration of the federal government converted over 100 vehicles to dual fuel operation on CNG or LNG and gasoline. They found that performance dropped from 25 to 40 percent for zero to 60 acceleration times, and the operating range was only about 50 to 75 miles.

Fig. 1-7. Southern California Gas Company's fleet of compressed natural gas-powered vehicles.

In spite of such limitations, the vehicles showed an impressive difference in operating costs. Depending on the local price of natural gas at the time, driving costs were reduced as much as two-thirds. One hundred cubic feet of natural gas is roughly equivalent to a gallon of gasoline.

Thousands of vehicles, mostly city or utility fleets, have been converted to use natural gas in order to save money. Pacific Lighting Corporation of Los Angeles, the Southern California Gas Company (see Fig. 1-7). Iroquois Gas Company in New York, Cities Service Gas Company in Missouri, and Piedmont Gas Company, in North Carolina, are but a few of the utility companies that have converted their customer-service vehicles to run on natural gas.

Although the opportunities for the individual motorist to switch to natural gas are still very limited, the costs of installing a compressing station for natural gas refueling could be shared on a community basis. It's also possible that, if enough demand existed, a local gas company might offer refueling services to motorists who want to have their vehicles converted. This could open up a whole new market for natural gas.

Chapter 2

Propane as an Alternative Fuel

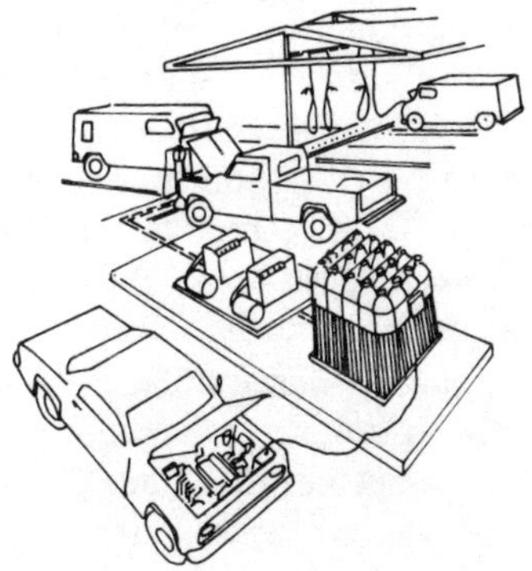

By now you should know that propane is a gaseous fuel that is normally stored in liquid form. You've also read that it is a clean-burning, high-octane fuel with good performance characteristics. But you may be wandering what exactly is that stuff that sounds so great?

Propane comes from beneath the surface of the earth. It was created millions of years ago through complex geologic processes that still aren't fully understood. Most of the propane is found in shallow natural gas wells and in the gas fields that surround oil wells. A stripping process is used to separate the propane from the other well gases (such as methane, butane, and ethane). The rest of the propane we use is a by-product of crude oil refining. The lighter the crude oil, the greater the amount of propane that's produced in refining.

Like gasoline and diesel fuel, propane is a *hydrocarbon*. A hydrocarbon is anything made from hydrogen and carbon. What distinguishes one fuel from another and gives each its set of physical properties is the relative number of hydrogen and carbon atoms per molecule. More carbon atoms ordinarily means the fuel will have a higher heat value.

One of the reasons why propane and gasoline contain different amounts of Btu (a measure of heat content) is because of the differences in their chemical structures. The chemical formula of propane is C_3H_8. This means each molecule of propane contains three carbon atoms and eight hydrogen atoms. Gasoline, which is a blend of several different hydrocarbons, contains compounds with formulas ranging from C_4 up through C_{14}. More carbon atoms means more heat value in the fuel. And that's exactly the way it is. Gasoline has more Btu per gallon than propane. Because there are more propane molecules in a pound of fuel than gasoline, propane comes out on top in terms of Btu per gallon. See Table 2-1.

It's not important to memorize or even understand all this chemistry. What's important is to realize the potential propane offers as an alternative fuel. Propane differs from conventional fuels in a number of ways. The most obvious difference is that propane is a gas. This means a propane-powered engine must have a special carburetor that can handle a gaseous fuel instead of a liquid fuel.

At ordinary temperatures and atmospheric pressure, gasoline and diesel fuel are liquids. Such fuels don't burn well unless they are first vaporized and mixed with oxygen. With gasoline, either a carburetor or fuel-injection system must be used to create a fine mist of liquid fuel. The mist is then mixed with air as it enters the engine. The tiny droplets of gasoline pick up heat from the surrounding air and from the engine. This helps the fuel to vaporize and mix with the air so that it will burn completely. Otherwise fuel economy, performance, and emissions will suffer.

When an engine is first started (especially on a cold morning), there isn't enough heat yet available to completely vaporize the gasoline. This results in incomplete combustion, inefficient use of fuel, and higher exhaust emissions until the engine warms up to its normal operating temperature. If the choke isn't working properly, the engine might hesitate, stumble, and stall until it warms up.

With diesels, the situation is somewhat different. Diesel fuel is mixed with air by injecting it directly into the combustion chamber as a highly pressured mist. The fuel isn't injected, however, until the air inside the combustion chamber has been compressed and is extremely hot (around 1000°F). The instant the diesel fuel hits the hot air it ignites. Because there is little time for the fuel to mix with the air, combustion is not complete. This explains why diesels typically emit a lot of soot and other pollutants in their exhaust. Injector timing and the injector's spray pattern must also be precise for the fuel to burn properly because there are no spark plugs to ignite the fuel mixture. Add this to diesel fuel's low volatility and you can see why diesel's have a reputation for being hard to start in cold weather.

Table 2-1. Comparison of Propane and Gasoline.

	Propane	Gasoline
Physical state during storage	Liquid	Liquid
Physical state when entering carburetor	Gas	Liquid
Boiling temperature at normal pressure	−44°F	80° to 450°F
Heat value in Btu/gallon	91,000	115,400 (average)
Heat value in Btu/pound	19,750	18,920 (average)
Octane ratings: Research Motor Pump	 112 97 104	 90 to 96 84 to 88 87 to 92
Exhaust emissions	Naturally low HC & CO	High HC & CO (requires special pollution controls)
Additives needed	None	Octane improvers such as tetraethyl lead, MMT, etc. Deicers Detergents
Weight (lbs./gallon)	4.86	6.8
Chemical formula	C_3H_8	C_4H_x to $C_{14}H_x$ (blend of various hydrocarbons)

How does propane differ from gasoline and diesel fuel? Because propane is a gas at normal temperatures and atmospheric pressure, it is already "vaporized" so to speak. No additional heat is required and the fuel mixes readily with air. The result is nearly perfect combustion, very low exhaust emissions, and few cold-weather starting problems.

Because propane is a gas, the propane carburetor's job is far simplier than that of gasoline carburetor, fuel-injection system, or diesel-injection system. All the carburetor has to do is meter the right amount of propane necessary to achieve the optimum air/fuel ratio. Because of its relatively simple function, a propane carburetor is usually called a *mixer*.

STORED AS A LIQUID

Propane has another interesting property. When compressed it turns into a liquid. Propane has a boiling point of about −44 degrees Fahrenheit. Unless it is kept under pressure at temperatures above −44 degrees, it will remain a gas. Below −44 degrees Fahrenheit, it will remain a liquid at normal atmospheric pressure.

Compressing propane raises its boiling point. The higher the temperature the more the pressure that is required to keep it from boiling. Release the pressure and it turns back into a gas. For example, about 13 psi is necessary to keep propane liquified when the temperature is 20° below zero. But at 100°F, 190 psi would be necessary to keep it from boiling. Most propane fuel tanks are designed to handle working pressures of 175 psi to 240 psi.

The reason for storing propane as a liquid rather than a gas is to save space. See Fig. 2-1. A given quantity of propane in liquid form would occupy roughly 270 times as much volume if allowed to expand into a gas. By squeezing the gas until it becomes a liquid, and then keeping the liquid under pressure, a far greater quantity of fuel can be stored inside a given fuel tank.

To get the propane back into the gaseous state so that it can be mixed with air and burned inside the engine, all that's necessary is to release the pressure. The pressure inside the fuel tank pushes the fuel through a fuel shut-off valve (called a *fuelock*), through the

Fig. 2-1. Propane is stored as a pressurized liquid. Working pressures are usually between 175 and 240 psi.

pressurized fuel line, and to a device called a *converter-regulator*. Here the high pressure liquid is allowed to expand until it becomes a low-pressure gas. Because the expanding fuel absorbs heat as it changes from a liquid back into a gas, hot water from the engine's cooling system is circulated through passageways in the regulator-converter to keep the unit from frosting up. The low-pressure propane gas is then drawn from the regulator-converter to the mixer. There it is metered into the incoming air and burned inside the engine. See Fig. 2-2.

APPLICATIONS

This clean-burning, efficient gaseous fuel you've been reading about can be used in one of three basic ways: as a straight replacement for gasoline in any spark ignition engine; as an alternative for gasoline in a dual-fuel conversion; or as a diesel fuel supplement in a diesel engine.

In a straight propane conversion, the gasoline carburetor (or fuel injection system), fuel pump, fuel lines and fuel tank are replaced by a propane mixer, regulator-converter, high-pressure fuel lines, a fuelock and pressurized fuel tank. The conversion is an excellent choice for many fleet applications because all dependence on gasoline is eliminated. This cuts operating costs and increases the payback savings more quickly. The total conversion approach

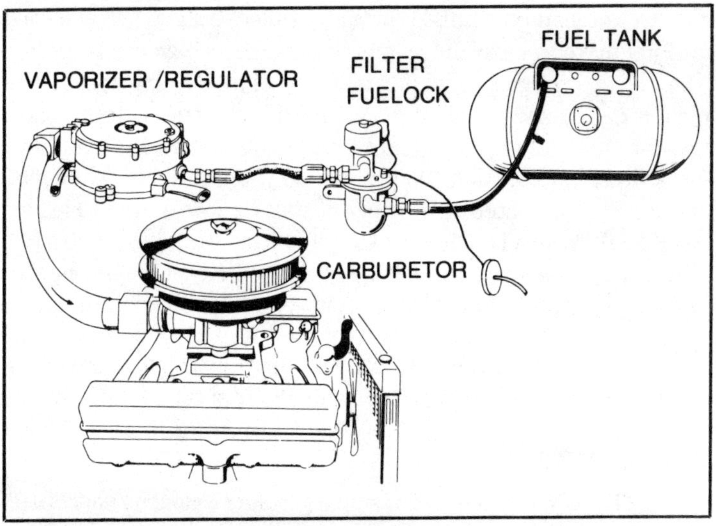

Fig. 2-2. Basic propane fuel system consisting of a pressurized fuel tank, fuelock, converter-regulator, and propane carburetor.

works best for short haul and local delivery fleets that return to a central garage nightly for refueling. Straight propane would only work for high-mileage vehicles where adequate fuel supplies were readily available.

In the dual-fuel conversion, the original gasoline fuel system is retained so that the vehicle can run on either gasoline or propane. The propane mixer is installed inside a special air cleaner housing atop the gasoline carburetor. Fuel shut off solenoids controlled by a dash switch allow the driver to select one fuel or the other. This is a practical alternative for many recreational vehicles, pickups, and private automobiles as it eliminates the worry over finding propane in unfamiliar areas. Gasoline can always be used as a backup fuel until a propane dealer can be found. The only drawback to this approach is that propane and gasoline have different performance characteristics. Tuning the engine to one fuel compromises mileage and performance when the engine is running on the other fuel. The solution is to tune the engine so that it runs best on the fuel that is used most often.

Another use for propane is as a diesel-fuel supplement. This would be a good choice for many farm tractors, construction equipment, and heavy-duty long haul trucks. Propane gas is fed into the diesel air intake system in metered amounts to reduce the amount of diesel fuel consumed by the engine. Fuel savings of up to 30 percent have been reported using this technique.

Unfortunately, straight propane (like straight gasoline and alcohol) makes a lousy diesel fuel because of its high octane rating. The octane rating of a fuel is a measure of its burning speed. The higher the octane rating the slower the fuel burns. In a spark-ignition engine, a high octane rating is desirable because it allows a fuel to resist detonation and preignition (the spark knock, pinging and run-on associated with many of today's low octane unleaded gasolines). A diesel engine, on the other hand, runs best with a fuel that ignites easily and burns quickly. A good diesel fuel is one that has a high *cetane rating* (the opposite of a high octane rating).

Propane has been successfully used to fuel stationary diesel engines, but such engines typically require very high compression ratios to offset the fuel's high octane and low cetane ratings.

OCTANE RATINGS

As far as using propane inside a gasoline engine is concerned, propane is an ideal substitute because its octane rating is higher than that of premium gasoline. The subject of octane ratings can be very confusing. There are three different ways to rate a fuel's resistance to detonation: ***research octane, motor octane*** and ***pump octane***.

The research octane rating is determined in a laboratory by comparing a test fuel's detonation resistance to that of two known fuels: iso-octane and heptane. Iso-octane is fairly resistant to detonation while heptane is not. The test fuel is compared to various mixtures of the two known fuels until a blend is found that detonates at the same compression ratio as the test fuel. If the test fuel detonates at the same compression ratio as a blend of 90 percent iso-octane and 10 percent heptane, for example, the fuel is given a research octane rating of 90. If the test fuel detonates at the same point as a blend of 99 percent iso-octane, it has a rating of 99. Should the test fuel be more resistant to detonation than iso-octane, it would have a number over 100.

Until recently, the research octane number was the one usually quoted because it is the highest of the three. For gasoline, the research octane is usually between 90 and 96. For pure propane the rating is 112. The problem with research numbers, however, is that they are based on ideal circumstances.

Motor octane ratings came about as a means to more accurately describe a fuel's resistance to detonation inside a typical engine. In the motor octane test, the known and test fuels are compared inside an engine that simulates real world conditions. The result is lower octane numbers. A fuel that has a 98-research-octane rating will typically have about a 90-motor-octane rating. The motor rating for gasoline is about 84 to 88 (depending on the grade of fuel and the type of additives used).

Tetraethyl lead is an excellent octane improver for gasoline, but the Environmental Protection Agency is in the process of eliminating lead as a fuel additive because of its harmful effects on the environment. Lead will also render the catalyst inside a catalytic converter useless. Cars made in the United States since 1975 (all of which have catalytic converters) are designed to run on unleaded gasoline.

Unleaded gasoline usually has an octane rating one to two points lower than leaded regular. Propane, by comparison, has a motor octane rating of about 97. That means it is more resistant to detonation than today's gasolines and it does not require the use of potentially harmful additives.

Because research octane numbers are unrealistically high and motor ratings are disappointingly low, government officials—with the assistance of the oil companies—decided that an average of the two rating systems would be a satisfactory compromise. The octane number you see posted on a service station pump, therefore, is nothing more than an average of the research and motor ratings. For premium unleaded gasoline, the rating is usually around 92. For straight propane it is 104. For commercial quality propane that contains up to 10 percent or more other gases (such as butane), the pump octane will be somewhat lower—maybe 96 to 100.

The advantages of a high octane fuel are many. Because the fuel resists detonation, it will be less likely to produce annoying and potentially harmful *pinging* or *spark knock* inside an engine. Severe detonation can quickly ruin an engine if not corrected. The uncontrolled explosions inside the combustion chambers can crack rings, blow out head gaskets, knock holes through the tops of pistons, and hammer rod bearings into oblivion. Severe detonation is usually the result of too much compression or too much spark advance for a given fuel.

A high-octane fuel is also less prone to ***run-on*** or be prone to ***dieseling*** when the ignition is switched off. Some engines will continue to run and chug after the ignition is turned off because carbon deposits on the piston tops and valves are so hot that they spontaneously ignite the fuel. This is called ***preignition*** and the only cures are to install a fuel shut-off solenoid on the carburetor, cleaning the carbon deposits from the engine, or switching to a higher-octane fuel.

High-octane fuels also have another important advantage. They allow the use of higher engine compression ratios for improved performance and fuel economy. During the late 1960s, when high octane leaded premium gasoline was readily available, compression ratios as high as 11.5 to 1 were not uncommon in passenger cars. But with the elimination of lead from today's gasolines and a general lowering of octane ratings, current compression ratios are around 8.5 to 1. Performance and fuel economy suffer as a result.

Although switching to propane won't necessarily make an engine run better, it can eliminate detonation and preignition problems (provided the conversion is done properly). If the engine is being overhauled, high-compression pistons or cylinder heads can be installed to up the performance and fuel economy.

PERFORMANCE

With a simple "bolt-on" conversion with no internal engine modifications to take advantage of propane's high octane qualities, propane will perform as good as and often better than gasoline. Add some internal modifications and the performance will increase accordingly.

One of the reasons why propane often outperforms gasoline is because it is a gaseous fuel. It mixes with the air more easily and thoroughly for better combustion. A gaseous fuel usually promotes better cylinder-to-cylinder fuel distribution. In a V-8, for instance, the four cylinders that are farthest from the carburetor typically receive less fuel than those closest to the carburetor. The lean cylinders produce less power than the rich ones, and the resulting power imbalance hurts performance. Turns in the intake manifold plumbing can also cause droplets of liquid fuel to separate from the fuel mixture. This creates rich and lean regions that also hamper combustion efficiency. A gaseous fuel such as propane tends to remain evenly mixed with the air so that every cylinder gets a balanced charge of fuel.

Gaseous fuels also perform better when it comes to cold-morning starts, and while the engine is warming up to its normal operating temperature. On a cold morning, gasoline can puddle in the intake manifold because there isn't enough heat to vaporize the fuel. If the carburetor choke isn't doing its job and creating an over-rich mixture to compensate for the unvaporized gasoline, the engine will stumble and stall until it warms up. This wastes fuel and hinders driveability. With propane there are no such problems.

Engines converted to propane often produce more horsepower than they did on gasoline because the fuel mixer is adjusted to deliver a proportionately richer air/fuel mixture to the engine. Although a proportionately richer mixture isn't necessary, many users of propane prefer to take advantage of the fuel's clean-burning qualities to tune the engine for more power.

Every fuel has its optimum air/fuel ratio. Engineers call the amount of air necessary to completely burn a given amount of fuel the *stoichiometric ratio*. This is the air/fuel mixture that would result in total combustion with zero pollutants left over under ideal circumstances. A richer mixture would produce more power, up to a certain point, and more exhaust emissions. A leaner mixture would produce less power and fewer emissions, up to a point, and then the mixture becomes too lean to ignite. This results in *lean misfire* that increases emissions (because the fuel isn't burned) and risks causing preignition and detonation.

For gasoline, the ideal or stoichiometric ratio is about 14.7 pounds of air to every pound of fuel. For propane, the ratio is about

15.5 pounds of air for every pound of propane. Gasoline averages about 18,920 Btu of heat energy per pound and propane averages about 19,750 Btu per pound. Propane contains more Btu per pound than gasoline, but it also requires more air (a leaner mixture) to completely burn the fuel.

What all this means is that theoretically the exact same amount of horsepower would be produced by an engine running on a stoichiometric mixture of either propane or gasoline. The propane's slightly higher Btu content would be offset by its slightly leaner air/fuel ratio. Of course this is all in theory. In the real world, the propane-powered engine would produce somewhat more horsepower at the stoichiometric ratio because of its improved cylinder-to-cylinder fuel distribution and more thorough mixing with air. In the real world few carburetors are set at the stoichiometric ratio. Most are calibrated somewhat richer to avoid lean-misfire problems and driveability problems that can result from lean mixtures.

FUEL ECONOMY

Although propane contains more Btu per pound than gasoline and produces more horsepower, a gallon of propane weighs less than a gallon of gasoline. In terms of energy content per gallon of fuel, gasoline comes out on top (115,400 Btu per gallon versus 82,870 for propane). Fuel consumption in terms of *miles per gallon* usually increases about 10 percent when you convert to propane. This is for a simple bolt-on conversion. Internal modifications that take advantage of propane's higher octane rating can be used to improve fuel mileage. Such modifications really aren't necessary when you consider the price difference between propane and gasoline. Even with a 10 percent loss in fuel mileage, propane's cost advantage more than makes up the difference.

In many applications, the drop in mileage is less than 10 percent. In some cases, there is no drop at all or even a slight improvement in mpg. Much depends on the type of engine being converted and how the vehicle is used. Vehicles that are used for short-trips, stop-and-go city driving often show a mileage improvement because their engine's are rarely run long enough to reach normal operating temperature. Because propane performs well under such conditions, a noticeable improvement in mpg usually results.

EMISSIONS

The Environmental Protection Agency keeps a close watch over our nation's air quality, and what effect various alternative fuels have on air pollution. Unlike gasoline, propane has a long-standing reputation for being a clean-burning fuel. The low-pollution characteristics of propane have long been appreciated by operators of fork lifts and other industrial vehicles that are used indoors.

Research conducted by the U.S. Bureau of Mines indicated that carbon monoxide (CO) emissions from engines using propane were 50 to 92 percent lower than emissions from the same engine when run on gasoline. Unburned hydrocarbons (HC) emissions using propane were found to be 38 to 62 percent lower than those for gasoline. Another plus is that the kinds of unburned hydrocarbons that come out of a propane-powered engine do not react with sunlight as readily as those produced by gasoline to produce smog.

Propane's clean-burning qualities can be readily demonstrated on an infrared exhaust analyzer. These machines use a small tail-pipe probe and infrared scanner to measure the amount of carbon monoxide and unburned hydrocarbons in a vehicle's exhaust. Carbon monoxide is a very dangerous pollutant. It can cause headaches, drowsiness, and nausea if breathed in quantity. It can also asphysiate a person if the concentration is high enough.

An average truck engine burning gasoline, for example, might register a CO exhaust content of 5 to 6 percent. The same engine on propane would probably show only 1 to 1½ percent CO.

When an engine is converted to propane, the technician will usually adjust the carburetor so that the idle CO is around 1 percent, and the cruise CO is around 0.2 percent. Although the carburetor could be adjusted to practically eliminate all CO from the exhaust, engine performance would drop noticeably. Therefore, a 1 percent CO level is a good compromise between low emissions and good performance.

Another area where propane comes out on top is with evaporative emissions. Vapors normally seep out of the fuel tank and carburetor when a vehicle is not running. Up to 35 percent of the pollutants given off by a gasoline-powered vehicle can be evaporative emissions. Since 1975, most of these emissions have been eliminated by the use of charcoal canisters and various plumbing in automobiles. Trucks, however, can still be gross polluters. Propane doesn't produce much evaporative emissions because the fuel system is sealed.

ENGINE LIFE

Propane's clean-burning characteristics means there's no buildup of carbon, varnish, or sludge inside the engine. Less crud in the crankcase means oil drain intervals and filter life can be safely extended for added savings. See Fig. 2-3.

Because the oil remains relatively clean, the engine lasts much longer than it would on gasoline. Oil is the engine's lifeblood. When sludge and acids build up as a result of combustion blow-by, the additives in the oil are rapidly used up. As a result, the bearings, rings, valve guides, and other friction surfaces begin to wear. But because propane burns much cleaner than gasoline—even while the engine is warming up—the amount of contaminants that find their way into the oil are greatly reduced. See Fig. 2-4.

Using propane will not harm a catalytic converter or damage an engine—with one exception. Propane does not contain lead or any other fuel additives. In engines designed to run on leaded gasoline, the lead acts as a lubricant for the exhaust valves to reduce wear. Eliminating the lead from the engine—either by converting to propane or running on unleaded gasoline—causes the exhaust valves to wear two to three times faster than normal.

To understand this process, consider the environment in which an exhaust valve operates. It is exposed to temperatures well over 1000°F—sometimes twice that—and it is hammered open and shut hundreds of times a minute. At 4000 rpm, for example, the exhaust valve is opening and closing 33 times a second. And every time it slams shut, a microscopic "welding" takes place between the steel in the valve face and the steel in the valve seat. When the valve opens, some of the seat metal pulls away and sticks to the valve face. The next time the valve slams shut, these tiny particles of metal are embedded in the valve face where they act like sandpaper to wear away the valve seat. Repeat this cycle thousands of times

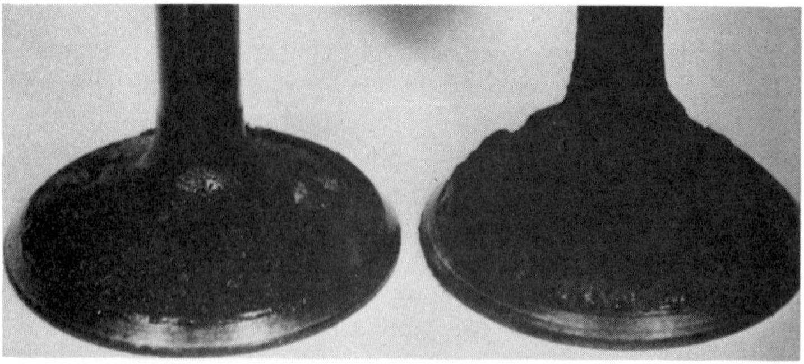

Fig. 2-3. Propane's clean-burning characteristics can be shown by these engine valves. The one on the left is from a propane-powered engine with 92,000 miles on it. The one on the right (caked with carbon deposits) is from an identical engine with 92,000 miles on gasoline.

Fig. 2-4. Reduced oil dilution allows propane-powered engines to go more miles between oil changes. Engine life is extended to as much as 200,000 miles.

over and over and the valve seal will eventually fail. High temperatures and high rpm operation only aggravate the situation.

In an engine designed to run on leaded gasoline, the lead forms a deposit on the wear surface between the valve face and seat. This helps to "insulate" the metals from one another so less welding and wear takes place. Eliminate this protection with a clean-burning fuel like propane and the wear begins to accelerate. This is especially true for vehicles subjected to hard use. In light-duty vehicles that don't run for long periods of time or that are subjected to high rpm use, valve wear is seldom a problem anyway. In such applications, switching to propane produces no problems.

Another factor to consider with older engines or those that are designed to run on leaded gasoline is the type of oil used. When a vehicle is converted to run on propane, some mechanics claim valve wear can be minimized by using a high ash content oil (do not use a synthetic oil). The ash works its way past the valve guides and rings to form a protective deposit between the valve face and seat. Although it isn't as effective as lead, it can reduce the rate of wear.

Specially formulated motor oils for propane conversions can also help reduce valve wear (Fig. 2-5). The calcium and barium additives that are used in regular oils, to disperse acids and varnish, can contribute to exhaust valve failure in converted engines because of propane's high combustion temperatures. The calcium and barium additives can form a film on the face of an exhaust valve and reduce heat transfer from the valve to its seat. This causes the valve to run hot and eventually burn. The specialty oils don't have such additives so longer valve life can be expected when they are used. The specialty oils also contain other additives that keep the oil from thickening or forming sludge deposits when extended drain intervals are used.

The best solution to preventing valve wear in leaded gasoline

Fig. 2-5. Specially formulated motor oils are available for propane-powered engines to extend valve life and reduce oil thickening when extended oil-drain intervals are used.

engines that are converted to propane is to install stainless steel valves or hardened stellite valve seat inserts. Another solution is hard-face inlay welding that creates a wear-resistant valve seat surface. Extra attention should also be given to grinding the valve seat. A seat width of 1/16 of an inch is all right for normal use of 3/32 of an inch for heavy-duty or turbocharged applications.

For engines designed to run on unleaded gasoline, and that includes every passenger car built in the United States since 1975, converting to propane poses no valve-wear problems because these engines already have the hardened valve seats. The seats are *induction hardened* at the factory to make them wear-resistant. Induction hardening is a metallurgical process that changes the structure of a metal to make it harder. See Fig. 2-6.

The absence of lead in propane also means that spark plugs will last as much as 40,000 miles before replacement is necessary. That is about four times longer than the plugs in an engine that runs on leaded gasoline, and somewhat longer than those in engines fueled by unleaded gasoline. The lead in gasoline forms deposits on the spark plug electrodes and that eventually causes the plugs to misfire.

REFUELING

Refueling a propane-powered vehicle is as easy and simple as refilling an ordinary gas tank. The only difference is that the fuel is under pressure. This requires a special type of nozzle connection and dispenser. The propane fuel tank is filled to only 80 percent capacity to allow for expansion. For example, if the tank is filled in the morning when the outside temperature is relatively cool, the fuel would need room to expand as the day grew warmer. If some room for expansion were not provided inside the tank, the build up of pressure would cause a safety relief valve to vent some of the excess propane out of the tank. That would be a waste of fuel and a possible safety hazard if the vehicle were parked indoors.

SAFETY

When treated with the same respect and common sense as any other fuel, propane is actually safer than gasoline or diesel fuel according to the U.S. Department of Transportation. Propane does not "puddle" when spilled. It dissipates quickly into the air because it becomes a gas the instant it escapes from the pressurized fuel system. The *flammability limits,* the conditions during which the right proportions of air and fuel are mixed to support combustion, are very narrow.

According to the National LP-Gas Association, propane engine fuel has established an excellent safety record over the past 60 years. National safety standards apply to all the components of the fuel system. These standards are kept up-to-date by the equipment manufacturers, the insurance industry, and government regulators. See Fig. 2-7.

The National Fire Protection Association issues these standards in its publication, *NFPA Standard #58; Storage and Handling*

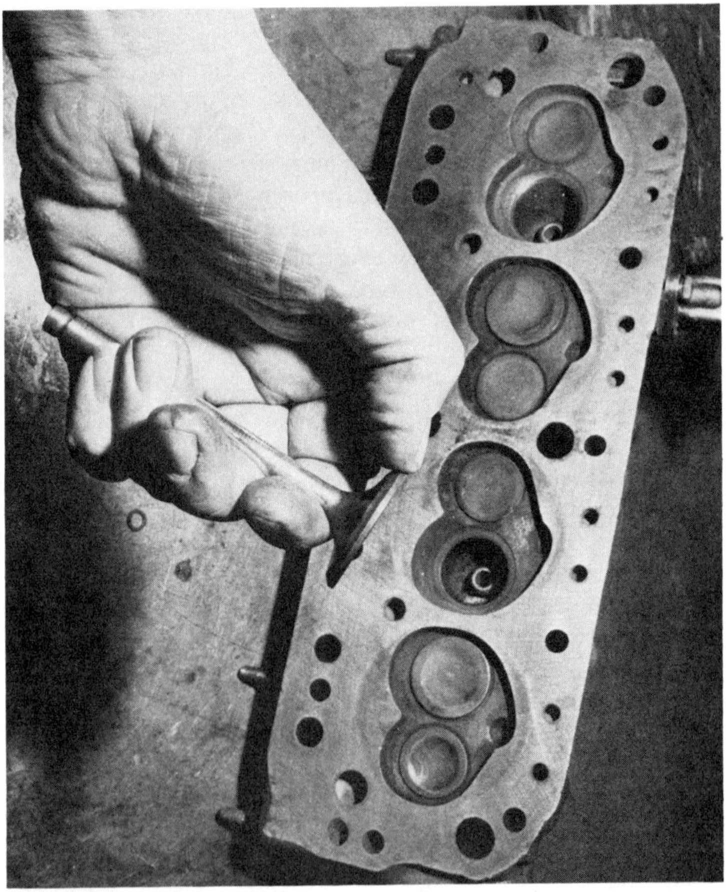

Fig. 2-6. Propane-fueled engines should have hardened valves and valve seats to reduce valve wear if the engine is to be subjected to hard use. Otherwise, wear of the exhaust valves can increase quickly in engines designed to run on leaded gasoline. There should be no difference in valve wear in engines designed for unleaded gasoline.

Fig. 2-7. Alternate-fuel powered vehicles have excellent safety records.

of Liquified Petroleum Gases. The same safety rules are issued by the American National Standards Institute. Most of the states have adopted these standards as their regulations applying to propane. Because of this, only *approved* components should ever be used in a propane conversion, and all parts should be installed in accordance with the rules.

Approved fuel tanks, for instance, are constructed of high-tensile-strength metal and are capable of withstanding pressures of 1000 pounds per square inch. Actual working pressures for propane storage range from 175 to 200 psi—a safety factor of five. In the event of a collision, a propane fuel tank has over 20 times the resistance to rupture of an ordinary gasoline or diesel fuel tank.

Tests were conducted by the Research Institute for Road Vehicles in the Netherlands to determine the safety of propane fuel systems by smashing vehicles into brick walls. Even when the vehicles were totally destroyed, the propane tanks remained intact. In a real accident, a propane-powered Volvo was struck by two freight trains traveling in opposite directions at a crossing. All that was left of the Volvo was its propane fuel tank!

Propane fuel systems are also fitted with safety valves that automatically shut off the flow of fuel in the event of a rupture in the fuel line. All fuel flow is also shut off when the ignition is switched off.

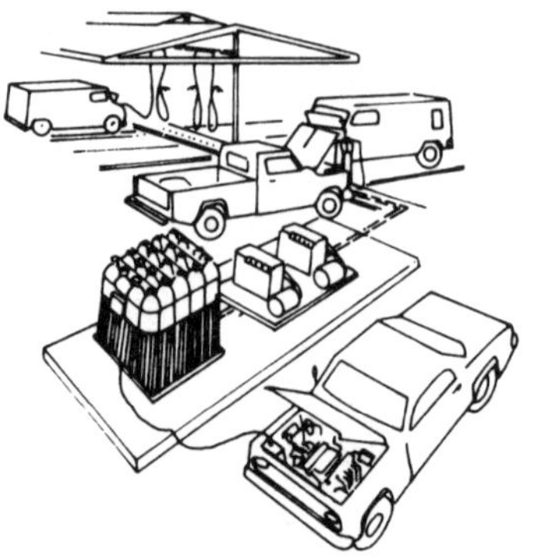

Chapter 3

Propane's Advantages

In this chapter, I will take a look at the many advantages of using propane as an alternative engine fuel. These claims are based on information from the National LP-Gas Association, various propane equipment manufacturers and propane suppliers, and most importantly, the actual experiences of fleet users.

No book about alternative fuels would be complete without an honest evaluation of the disadvantages as well as the advantages. Fortunately, propane doesn't have many drawbacks. But there are some disadvantages that can have an important bearing on whether or not propane is a suitable alternative for a given application. These points are covered in the next chapter. Be sure to read this chapter and the following one before you decide whether or not propane is for you.

LOWER FUEL COSTS

The main reason for the big rush to propane in recent years is its price advantage over gasoline. Propane sells for roughly one-third less per gallon than gasoline in most parts of the country. In some areas of the world, it is extremely inexpensive.

In late 1981, propane was selling for 40 cents to 60 cents a gallon less than unleaded gasoline; the average price advantage was about 50 cents a gallon. In Table 3-1, the annual fuel savings based on 40-cent, 50-cent, and 60-cent price advantages are calculated for various applications.

Table 3-1. Fuel Savings with Propane.

Annual savings with propane selling for 40¢ a gallon less than gasoline

mpg*	5,000	10,000	15,000	20,000	25,000	30,000
			miles driven per year			
5	$360	720	1080	1440	1800	2160
7.5	$239	479	720	958	1197	1581
10	$180	360	540	720	900	1080
12.5	$144	288	432	576	720	855
15	$120	239	360	479	599	718
17.5	$103	200	309	411	514	617
20	$ 90	180	270	360	450	540

Annual savings with propane selling for 50¢ a gallon less than gasoline

mpg*	5,000	10,000	15,000	20,000	25,000	30,000
			miles driven per year			
5	$450	900	1350	1800	2250	2700
7.5	$300	599	900	1199	1499	1800
10	$225	450	675	900	1125	1350
12.5	$180	360	540	720	900	1080
15	$149	300	450	600	749	900
17.5	$128	257	385	514	643	772
20	$113	225	337	450	563	675

Annual savings with propane selling for 60¢ a gallon less than gasoline

mpg*	5,000	10,000	15,000	20,000	25,000	30,000
			miles driven per year			
5	$540	1080	1620	2160	2700	3240
7.5	$360	720	1080	1440	1800	2160
10	$270	540	810	1080	1350	1620
12.5	$216	432	648	864	1080	1296
15	$180	360	540	720	900	1080
17.5	$154	309	463	618	771	927
20	$135	270	405	540	675	810

* Average mileage on propane fuel is about 10% less than that on gasoline. This has been included in the annual savings calculations. Conversion costs are not included.

To estimate the amount of money you'd save each year by switching to propane, find the price difference in Table 3-1 that most closely approximates the retail prices in your area. Then find your vehicle's average miles-per-gallon figure and read across under the appropriate column for the number of miles you drive that vehicle each year. This is how much you'll save in yearly fuel bills by converting to propane.

For example, say that you have a fleet of light delivery trucks (Fig. 3-1) that average about 10 mpg around town. Each truck is driven about 25,000 miles a year. If propane is selling for 60 cents a gallon less than gasoline in your area, you'll save $1350 per year on each vehicle if you switch to propane. If it costs $800 to convert each truck, the conversion will pay for itself in less than a year and you'll still save $550 in fuel costs. With each successive year, the savings will probably increase because the price of gasoline is headed up.

For RV owners, the numbers are equally impressive. Say that you own a big motorhome that gets approximately 5 mpg. During the vacation season, you travel about 10,000 miles. If you convert your motorhome to propane, you'll save $1080 per year if propane sells for 60 cents a gallon less than gasoline. That's enough to pay for the cost of converting to propane the first season. The following year, you'll reap the full savings that propane offers.

What about the salesperson who drives an Oldsmobile Cutlass? If the car gets 15 mpg on the road and is driven 25,000 miles a year, savings will be $900 the first year with a 60-cent price advantage for propane. That's not quite enough to break even the first year, but the following year the savings start to add up. The cost of conversion would be mostly paid back the first year.

Even if the conversion costs (they range from $800 to $1400 or more) are not offset the first year, the fuel savings will likely make up the difference before the end of the second year. The payback period depends on the vehicle's mpg and the number of miles driven annually.

SURPLUS OF FUEL

Although supplies of gasoline have increased recently, there are still lingering high fuel prices from the shortages of the mid-1970s. What's more, we are still very dependent on the politically unstable Middle East for a large percentage of our oil. Should something happen in that part of the world to disrupt the flow of oil, we'd be in for another period of severe shortages and perhaps rationing.

With propane, the possibility of shortages is remote. Most of the propane we use in this country is produced right here in the United States. Although "experts" say that natural gas supplies have been gradually declining, recent finds seem to have reversed the trend. Canada and Mexico also have large supplies of natural gas from which propane can be extracted. On the world scene, there is a virtual glut of propane; prices are as low as a few cents a gallon in some of the oil-producing countries.

Another factor to keep in mind is that the demand for propane as a motor fuel is nowhere near the demand for gasoline. The number of propane-powered vehicles in this country is still a very small percentage of the total 130,000,000 vehicles on the road.

Fig. 3-1. The Los Angeles *Times* has racked up millions of miles on its fleet of propane-powered delivery trucks. Many of the vehicles went 200,000 before requiring any major repairs.

Even if every fleet, motorhome, and pickup truck were converted, there would be adequate supplies of propane available.

FUEL AVAILABILITY

According to the National LP-Gas Association, there are about 25,000 propane dealers nationwide. They are most often listed in telephone directories under "Gas, Liquified Petroleum—Bottled and Bulk."

Fleet operators usually find it most advantageous to have a storage tank installed right on their premises. These are usually sold or leased to the operator, and they consist of a tank, a pump, and a meter. Many truck stops also carry propane fuel. Any place that fills propane bottles used in campers, trailers and recreational vehicles could fill your fuel tank if they are equipped to report road taxes on motor fuel. Compared to other alternative fuels such as compressed natural gas, alcohol, or hydrogen, propane is readily available.

LOWER FUEL TAXES

Many states have reduced or, in some cases, eliminated road-uses taxes on sales of propane motor fuel. This is to encourage the conversion of vehicles to propane so that additional gasoline supplies will be available for other motorists.

SIMPLE CONVERSION

In most cases, converting to propane does not require major engine modifications. The propane fuel system is all bolt-on equipment. Most competent mechanics can complete a conversion in six to seven hours or less.

For a straight conversion, the existing gasoline carburetor, the fuel pump, the fuel lines, and the gas tank are replaced by a pressurized propane fuel tank, a fuelock/filter, pressurized fuel lines, a converter-regulator, and a propane carburetor. In dual fuel conversions, the original gasoline fuel system is retained and the propane fuel system is added to it. The propane carburetor is mounted on top of the gasoline carburetor. With the flick of a switch, the engine can be run on either gasoline or propane.

After the propane fuel system components are installed, the vehicle is tuned for optimum performance on propane. This involves adjusting the carburetor mixture, ignition timing, and spark advance curve.

RECONVERTING TO GASOLINE

Because the original gasoline fuel system components are not modified, they can be stored for future use if necessary. When the vehicle is sold, for example, the original owner might want to keep the propane fuel system for use in his next vehicle. Reconverting to gasoline is as simple as removing the propane parts, reinstalling the gasoline parts, and retuning the engine to its original specifications. Reconverting to gasoline often doesn't take much time because when the original gasoline fuel tank and lines are left in place they do not have to be reinstalled.

PROVEN TECHNOLOGY

With any alternative fuel, there is always a concern over whether or not the fuel really works. Propane has been used to power engines almost as long as gasoline. Since 1938, when official records were first kept on the use of LP-gas, more than 28 *billion* gallons of propane have been burned as motor fuel in the United States (according to the National LP-Gas Association).

As for the equipment, the propane fuel system component manufacturers represent a well-established industry. Since 1970, over 2.8 million LP-gas carburetors have been sold for use on both vehicle and stationary engines.

RELIABLE FUEL SYSTEM

The carburetor on a propane-powered engine is a simple device with few moving parts. There is no choke to stick and cause starting problems. There are no floats to cause flooding or fuel starvation. There's not even a fuel bowl to cause vapor lock. In other words, the propane carburetor is a relatively trouble-free device that just does its job without a lot of complicated parts and circuits.

As for the rest of the fuel system, no fuel pump is needed because propane is stored under pressure in the fuel tank. The electrically-operated fuelock prevents the flow of fuel when the ignition is off.

Compared to gasoline fuel injection or diesel engines, the propane fuel system is extremely reliable. Fleet users support such claims with years of experience and millions of miles of service. Some systems have gone 20 years or more in use with the only repair expense being a rebuilt vaporizer.

PERFORMANCE

Dynamometer testing has shown that performance on propane is as good as that with gasoline—sometimes better. The actual horsepower and torque output of a given engine is ordinarily within plus or minus 5 to 10 percent of its original ratings when converted to propane. Many users report propane improves performance because its high octane allows the use of more ignition advance and higher compression ratios.

Cold-Weather Performance

Because propane is a gas when it enters the carburetor, it mixes with air very easily. This improves combustion efficiency when the engine is cold and eliminates many of the problems associated with the usual "warm-up" period necessary with gasoline.

Until the engine warms up, an extra-rich mixture of gasoline must be fed to the engine. This is the choke's job. But with today's low-emissions, lean-calibration carburetors, an engine will often stall or hesitate until it reaches its normal operating temperature. Not so with propane.

As for cold-weather starting, propane vaporizes easily all the way down to −44 degrees Fahrenheit. Try starting a gasoline or diesel-powered engine at 20 below zero.

Mileage

Generally, there is not much difference in fuel mileage between propane and gasoline. On the average, most users report a 10 percent drop in mpg when converting to propane. Some users have found no difference or a slight improvement. Most engines get fewer miles-per-gallon with propane because propane contains fewer Btu per gallon than gasoline. For the engine to produce the same horsepower, therefore, more fuel is required.

Emissions

Propane typically produces lower amounts of carbon monoxide and unburned hydrocarbons than gasoline. Those hydrocarbons that it does produce are less reactive with the atmosphere. This helps to reduce the formation of smog and reduces the risks to those who are exposed to the exhaust. Nitrous oxide emissions are about the same and evaporative emissions are practically zero. Compared to a soot-belching diesel engine, propane is quite clean.

Thanks to its clean-burning characteristics, propane has long been used in industrial fork lift trucks inside large warehouses, factories, and mines. In areas with periodic motor vehicle exhaust inspection programs, switching to propane can help older, "dirty" engines pass government clean-air standards.

Longer Engine Life

According to the National LP-Gas Association, many fleet users can expect to last 180,000 to 200,000 miles on propane. The clean-burning characteristics of the fuel minimize oil contamination. This in turn helps to lengthen the life of rings and bearings. The high octane rating of the fuel also reduces the harmful effects of combustion knock. This reduces stress and shock loading of pistons, rings, connecting rods, the crankshaft, and related components so they last longer. Propane does not leave lead or varnish deposits on engine components and does not wash engine oil from the cylinder walls. All this adds up to less wear and tear on the engine for more miles between overhauls.

Trucks used for heavy work should have hard seats under the exhaust valves to minimize valve wear that can sometimes be a problem in engines designed for leaded gasoline. In many newer truck engines and every automotive engine built since 1975, valve wear is no different with propane than unleaded gasoline.

Longer Spark Plug Life

With propane, there is no buildup of combustion chamber deposits on spark plugs as there is with leaded gasoline. Spark plugs can last four or five times longer (up to 40,000 miles between changes). This helps to reduce maintenance costs for added savings. See Figs. 3-2 through 3-8.

Longer Oil Life

Propane does not contaminate or dilute the oil in the crankcase like gasoline does. This allows the oil to retain its lubricating qualities three to four times longer (up to 25,000 miles) between drain intervals. This also helps to reduce maintenance costs and prolong engine life.

Reduced Maintenance Costs

Longer plug life, longer oil drain intervals, and longer engine life all add up to significant maintenance savings for many propane users. Many fleet operators also report that the exhaust systems and mufflers on propane-powered vehicles last up to twice as long as they did on gasoline.

Less Downtime

For fleet users, reduced downtime is a very important plus because the time lost when a vehicle is down for repairs can often be

Fig. 3-2. Spark plugs last up to 40,000 miles thanks to propane's clean-burning characteristics.

Fig. 3-3. Believe it or not, this Chevy pickup can go 1800 miles on one fill-up of LP gas.

as costly as the actual repairs themselves. Longer engine life, longer oil life, and longer plug life with propane helps to keep a vehicle on the road more miles between repairs. The fuel system's reliability and the ease of cold-weather starting also reduces the amount of money spent on emergency service calls and breakdowns for many fleets.

Eliminates Pilferage

Because the propane fuel system is under pressure, it's impossible for anyone to steal the fuel by siphoning it out of the fuel tank. Unless the potential thief has a propane-powered vehicle to use the stolen fuel in, there is little incentive to "borrow" someone else's fuel.

Fig. 3-4. Here's how that Chevy pickup can go 1800 miles on a tankful of propane. It has a wall-to-wall fuel tank.

Fig. 3-5. Many fleets are converting to propane to realize significant fuel and maintenance savings. In Michigan, for example, the state is undertaking a program that could eventually lead to conversion of 13,000 school buses.

Some fleet operators have reported pilferage losses of as high as 10 percent or more. When they converted to propane, the losses disappeared completely. With a sealed fuel system, there is little or no loss due to evaporation or spillage while refueling.

Fig. 3-6. Propane-powered vehicles can be refueled from pumping stations such as this. It works just like an ordinary gasoline pump (showing gallons of fuel and dollar amount).

Fig. 3-7. The filler connections on all LP tanks for vehicles use a standard screw-on fitting.

FLEET EXPERIENCE

The following excerpts were taken from a speech delivered by George C. Blair, vice president and director of transportation for the Interstate Brands Corporation. The speech was made before the American Bakers Association September 14, 1980.

"We have been running LP in our route now for almost six years. We started out with 27 test vehicles. We gathered data over a period of about four and a half years, and I think we gathered very viable data that enabled us to put pencil to paper to look at the economic payback. We were somewhat shocked. We found it very hard to believe that here was something that has been around for 45 years. It's a fuel that has been used in Europe for many, many years. It's a fuel that's been used in this country in some major fleets, but no one has ever paid attention to it. So we started on the full-scale program (of conversion). Today, we have approximately 1000 of 4200 vehicles converted. We intend, because of economic reasons, to convert approximately 3800 vehicles to LP . . . .

"Our average differential in price between a gallon of propane and a gallon of gasoline will run about 44.6 cents, and that includes taxes. This will equate to somewhere between about 2.8 to 3.1 cents per mile less in fuel costs . . .

"Not only are there economics to be gained in the fuel itself, but longevity of the equipment. A 292 engine, which is our basic engine, we expect roughly 90,000 miles before we have to touch it. Burning LP, because it is so clean-burning, there is no gumming up, no varnish, etc., in the engine. We expect somewhere around 150,000 to 160,000 miles out of that engine. The exhaust system, the muffler, tailpipe, will virtually last the life of the chassis because, again, there are no contaminants going through it . . .

"Spark plug life--our new Preventative Maintenance program that we have just published for those vehicles converted to propane—calls for changing spark plugs at 45,000 miles. Engine oil—heretofore those running on gasoline were 6000 miles between oil changes. Now its 25,000 miles with LP . . .

"We have our lab set up to do oil analysis on our tractors for extended oil change intervals . . . and to date we have had no problems at all.

"I get awfully nervous when people say to me, 'Well what are the negatives of LP?' and I honestly have to say that we haven't found any negatives. We truly haven't. We plan, as I said, to convert about 3800 vehicles."

LONGEVITY RECORD

One of the most amazing examples of propane power was reported in the *Arkansas LP News.* A 1973 Ford F-250 pickup truck,

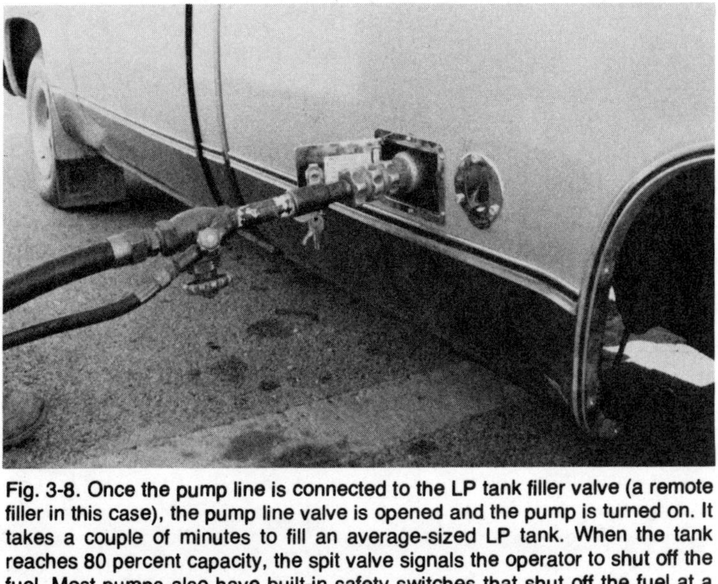

Fig. 3-8. Once the pump line is connected to the LP tank filler valve (a remote filler in this case), the pump line valve is opened and the pump is turned on. It takes a couple of minutes to fill an average-sized LP tank. When the tank reaches 80 percent capacity, the spit valve signals the operator to shut off the fuel. Most pumps also have built-in safety switches that shut off the fuel at a preset pressure to prevent over-filling.

with a 390-cubic-inch engine, logged 745,000 miles on the original engine while hauling loads of newspapers from Little Rock, Arkansas to Texarkana, Texas between January 6, 1973, and October 31, 1978.

A break-in period of 15,364 miles was driven on gasoline before the engine was converted to run on propane. An Impco Model CA425 carburetor, a Model E vaporizer-regulator, and a VFF30 fuelock/filter were installed by Liquigas, Inc., North Little Rock, a propane bulk plant.

At the time of conversion, owner Aubrey Davidson put the pickup into service hauling the *Arkansas Gazette.* The 165-mile, one-way distance was transversed loaded with an average of 4000 pounds of newspapers. Some loads were as much as 6000 pounds. The return trip was a deadhead run.

During the nearly six years the Ford hauled newspapers, about 25 different drivers were used. Bob Houston, vice president of Liquigas, said that during the period the 745,000 miles were logged, the truck put over 600,000 miles on *one set of spark plugs!* It also once ran 200,000 miles without a valve job. "These are astonishing mileage totals and an exceptional demonstration of engine life even with clean-burning propane," Houston comments.

The only major engine service was at 50,000 miles to repair a faulty oil pump. The rod bearings were replaced at this time because the oil pan had to be removed to get at the pump. At last report, the old Ford was still running strong. It has since been retired from hauling newspapers and is now used for everyday transportation by Aubrey Davidson. He says he plans to get at least another 100,000 miles on the same engine using LP gas.

Chapter 4

Propane's Disadvantages

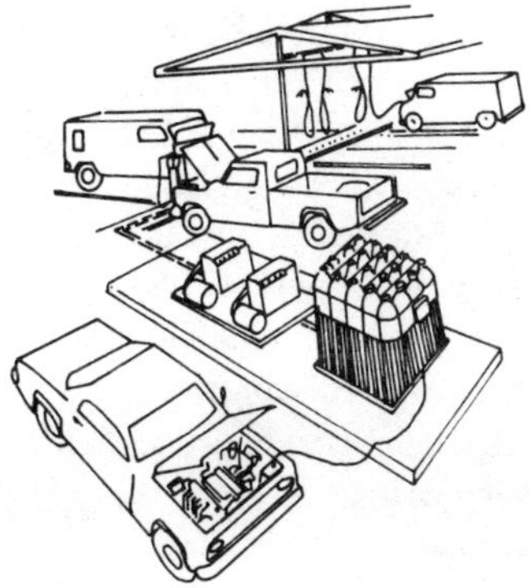

Although the advantages of using propane as an alternative fuel far outweigh the disadvantages for many applications, there are some drawbacks that must be considered.

HIGH CONVERSION COST

The number one reason why many people reject propane is because of the high initial conversion cost. Prices at most conversion centers range from $800 to as much as $1400 or more. The actual list price for the carburetor, converter-regulator, fuelock/filter, and related fuel line hardware is approximately $300. The price of the fuel tank depends on its size, but for an "average" number figure approximately $250. The rest of the cost is for installation labor. Converting passenger cars tends to be more costly than converting trucks because the trunk must be sealed off to prevent any possibility of propane seeping into the passenger compartment. Trucks are no problem because the fuel tank can be mounted almost anywhere outside the cab.

The average person who wants to convert his car, truck, or motorhome to propane will find that most equipment distributors won't sell parts to do-it-yourself mechanics. This is not because the conversion is difficult (it isn't), but out of fear that the novice mechanic might not do the installation properly, i.e., safely. In other words, the distributors want to avoid the possibility of any legal ramifications if something goes wrong with a do-it-yourself conversion. For Mr. and Mrs. Average Motorist, converting to propane means finding a local conversion center and paying them full fare for their services.

Having a conversion center do the work, however, is really the best way for most motorists to go. Although it costs more than a do-it-yourself job, the conversion center will do a professional installation. This means the parts will be installed properly (safely), the engine will be tuned for optimum performance (no guess work), and the work will be guaranteed.

For fleets, the initial conversion costs might be somewhat less if the fleet operator buys parts at a volume discount or direct from a distributor or manufacturer, if the fleet has a trained mechanic who can do the conversions, or if a local conversion center gives the fleet a special price because they're converting a number of vehicles. In such cases, the conversion costs may be as low as $500 to $700 per vehicle.

CONVERSION PAYBACK

To come out ahead with propane, you have to drive enough miles to offset the initial conversion expense (see Table 4-1). If you spend $1000 to have your vehicle converted, you won't start saving money until you have driven enough miles to pay for the cost of modifying your vehicle. If you have a truck that averages 12.5 mpg, and propane sells for 60 cents a gallon less than gasoline, the payback period will be almost 25,000 miles (see Table 3-1). If you drive that truck 25,000 miles a year, it will take you one year to offset the initial conversion expense. If you only drive 5000 miles a year, it will take you five years to reach the break-even point.

The high initial conversion cost and payback period is why many motorists don't even consider propane as a worthwhile alternative. They simply don't drive enough miles to realize any real savings. For Joe Commuter who drives 12,000 miles a year in a little economy car that gets 30 mpg, converting to propane doesn't make sense. Although he'd be saving money everytime he filled his tank, it would take him 55,000 miles to reach the break-even point. At 12,000 miles a year, that's slightly more than four years of driving.

That's why propane isn't for everyone. Unless a vehicle is driven a lot of miles or gets lousy gas mileage, converting to propane is an expensive way to save a few bucks at the gas pump. You have to figure your vehicle's average mpg and how many miles a year you drive before you can say whether or not propane can save you money.

Table 4-1. Conversion Payback Economics.

Example: The following numbers are based on a conversion cost of $1,000 and a 60¢ per gallon price advantage for propane.

Vehicle mpg	Break even point where fuel savings offset conversion costs*	
5	9,166	miles
7.5	13,750	miles
10	18,334	miles
12.5	22,916	miles
15	27,500	miles
17.5	32,084	miles
20	36,667	miles
22.5	41,251	miles
25	45,835	miles
27.5	50,417	miles
30	55,001	miles

*Mileage figures include 10 percent loss in mpg on propane.

For high mileage applications such as taxicabs, commercial delivery vehicles, police cars, garbage trucks, light utility trucks, traveling salesmen, long distance commuters and owners of big gas-gobbling motorhomes, propane is an ideal alternative. For Mrs. Housewife who only makes a couple of trips to the store each week or Mr. Average who puts less than 10,000 to 15,000 miles a year on the family sedan, converting to propane really isn't economical—unless there's serious concern over fuel availability.

If something happens in the Middle East to restrict the flow of oil to the United States, gasoline could be in short supply once again, and most likely with little warning. Under such circumstances, those with propane-powered vehicles would still be driving while others waited in line to buy gas. If they could get gas at all. For many businesses, being able to get from A to B on a regular schedule is an absolute necessity for survival. With no alternative to turn to, a severe gasoline crisis could put a lot of businesses out-of-business.

FUEL AVAILABILITY

There are about 25,000 propane dealers nationwide. Unfortunately, not all are equipped to refuel automotive propane tanks. By comparison, there are over 160,000 service stations that sell gasoline.

For the average motorist who converts to propane, buying the fuel might not be as convenient as buying gasoline. Except for many truck stops, few service stations sell propane. Many propane dealers are located in out-of-the-way locations, they are closed evenings, and many might or might not be open on weekends.

Traveling cross-country in a propane-powered vehicle could also be frustrating if the motorist had to search out the local propane dealer whenever he needed fuel. The National LP-Gas Association and State Associations all publish directories of propane dealers. Without a copy in your glovebox, you might find yourself spending some time searching for a place to buy fuel.

Perhaps this situation will change with time. A few years back, buyers of diesel-powered cars encountered the same kind of problems. Few service stations used to carry diesel fuel. About the only place a Mercedes owner could find fuel for his car was at a truck stop. That often meant driving miles out of his way just to fill his tank.

For the fleet user, propane availability is rarely a problem. Because most of the vehicles are returned to a central location nightly, they can be refueled from an on-site bulk storage tank.

LEGAL RESTRICTIONS

In spite of the fact that propane has as good a safety record as gasoline, propane-carrying vehicles are prohibited from using certain tunnels and bridges in some areas of the country. These laws are aimed primarily at the big tank trucks that haul large quantities of propane. In many instances, though, the laws are worded so that they apply to *any* vehicle that carries propane. If you want to get technical, and sometimes local law enforcement officials tend to get that way, such laws could apply to the propane bottles that are commonly found on travel trailers and campers.

The fuel tank on a propane-powered car or truck really doesn't carry enough fuel to pose much of a threat, and the chances of getting caught passing through a restricted tunnel or bridge are slim. Nevertheless, it could be against the law. Such laws are meant to protect the public. If a tank truck carrying a large supply of propane were to be involved in an accident in a tunnel, for example, a ruptured tank could turn the tunnel into an instant death trap either through fire or asphyxiation (propane is a gas, remember).

The odds of fuel tank explosion in a passenger car or truck accident are probably less than for gasoline. Tests have shown that heavy steel propane fuel tanks rarely fail. In severe accidents, the propane tank is often the only part left intact.

SAFETY

Many people are afraid of propane because of what they've heard on the news. "A derailed railroad tank car carrying propane exploded today and leveled half of Fairfax county." It happens but rarely in vehicles. There are several built-in safety features that prevent the loss of propane from the fuel tank in the event of an accident. What's more, the fuel tank itself can't explode because there's no oxygen inside it.

The biggest danger results from over-filling the tank in cold weather. If the vehicle is pulled inside a warm garage, pressure could build up inside the tank and cause some fuel to vent to the outside through a safety release valve. Ordinarily this causes no problems. If the vehicle is parked inside a closed garage and there happens to be an open flame nearby—such as the pilot light on a furnace or water heater—it's goodbye garage.

It is highly recommended that propane-fueled vehicles never be parked in any building that is not ventilated or contains an open flame source. Like gasoline, propane must be treated with respect.

Some insurance companies might be hesitant to insure a propane-powered vehicle, and they might ask for a higher premium. They might also require an inspection or certification of the fuel system components before they'll sell you coverage. Such details should be discussed with your insurance agent *before* you invest in converting to propane.

HANDLE WITH CARE

As with gasoline or any other flammable liquid, propane should always be handled with care. The air/fuel combustion limits for propane are between 2.5 and 9.5. In other words, if the relative proportions of air and propane are between these concentrations, the fuel will burn.

Because propane becomes a gas as soon as it escapes from a fuel line or fuel tank, a leak will usually dissipate before a dangerous concentration of propane can accumulate. Filling a propane fuel tank is no more dangerous than filling a gasoline tank. Actually it's probably safer.

Potentially dangerous situations can develop, however, if the escaping gas has no place to escape to. If fumes build up inside a sealed compartment, such as an unvented automobile trunk or a boat hull, an explosion could result if a spark set the mixture off. Providing adequate ventilation, therefore, is essential.

Propane is heavier than air. Escaping gas can cause problems for garage areas with sunken work pits. The gas can accumulate in the work pit and pose a real danger. The gas might displace enough oxygen to cause a person to pass out. Adequate ventilation must be provided and precautions must be taken to avoid trouble.

People must also avoid direct contact with propane liquid. If liquid propane is sprayed on bare skin, it can cause frostbite as it evaporates. When it turns from a liquid into a gas, it absorbs heat so rapidly that it can freeze skin.

SERVICE SNAGS

Propane fuel systems have a solid reputation for reliability, but nothing is perfect. A breakdown can cause you real problems if it happens someplace where there are no experienced propane mechanics available. For fleets, this isn't much of a problem because all of their vehicles are normally serviced in-house or at a local garage. For the owner of a propane-powered motorhome who finds himself stranded in Nowhere, North Dakota, finding someone who can fix his fuel system might be about as easy as cashing an out-of-town check.

The same applies to replacement parts. Because propane fuel systems are not that common, you can't get rebuild kits at your local Sears store or corner auto parts store. Most conversion centers have a good supply of parts on hand. If they don't have the exact replacement part you need, you might be in for a lengthy wait while the part is back ordered from the manufacturer. Of course, the same problem often confronts the owners of new cars or import cars when something breaks.

For fleets, having a reliable and ready source of replacement parts is extremely important to avoid unnecessary delays and downtime. So before you let anyone sell you a conversion, make sure they can back up their work with parts and service.

PERFORMANCE

Propane runs as good as gasoline in most vehicles, but it tends to be a less "forgiving" fuel. The ignition system, carburetor adjustments, and timing must be up to specifications for the engine to run properly on propane. There is also some danger in over-advancing the ignition timing. Unlike gasoline which will "knock" loudly if pushed beyond its octane limits, propane gives little warning. The result can be severe engine damage if the situation is neglected. If the engine is timed properly, however, there is no danger of detonation damage.

Propane has also been accused of causing *valve recession* problems in some engines. This is where the exhaust valves wear at an accelerated rate because of propane's clean-burning characteristics. Propane leaves no deposits behind to "lubricate" the valve seats. The result can be premature valve failure if the engine is subjected to hard use.

It must be emphasized that valve recession is only a problem in engines that are designed to burn *leaded* gasoline, and then only if the engine is run hard or at high speeds. Short-trip, city driving rarely causes valve wear problems in such engines because the valves don't get that hot. Valve recession is *not* a problem in engines with hardened valves and valve seats. This includes any engine that is designed to operate on *unleaded* gasoline. See Fig. 4-1. If valve recession appears to be a potential problem in a conversion, hardened valves and valve seats can be installed to minimize wear. This does add to the conversion expense.

Another area where propane has been accused of causing problems is that of *ring seating.* Because the fuel burns clean and does not wash the protective coating of oil from the piston cylinder walls, the piston rings might not receive enough wear to seal correctly. This can be a real problem in a brand new engine that is equipped with hard chrome steel rings. The problem can be prevented one of two ways: break the engine in on gasoline before converting to propane, or use standard cast iron piston rings instead of chrome steel.

SPACE LIMITATIONS

Propane tanks are fairly bulky (Fig. 4-2). On trucks they pose no problem because they can be mounted under the chassis. On the other hand, in cars the tanks are practically always mounted in the trunk because that's the only place where there's enough room. The tradeoff is a loss of some trunk space.

In many conversions, the tank or tanks can be mounted in the trunk space over the rear axle where the spare tire normally sits. This minimizes the loss of usable trunk space, but it creates the problem of where to put the spare.

Another limitation is that cars with hatchbacks are difficult to convert because there is no trunk. A box must be constructed around the fuel tank to seal it off from the rest of the passenger compartment. This is a necessary safety precaution to prevent

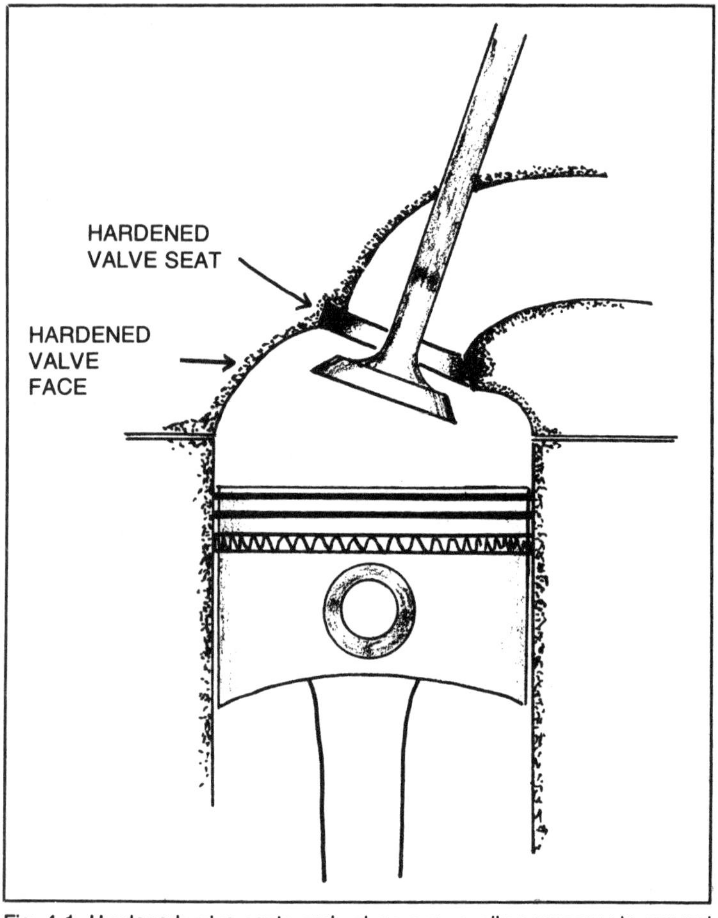

Fig. 4-1. Hardened valve seats and valves are usually necessary to prevent premature valve wear in propane-fueled engines. These parts are already included in engines designed to run on unleaded gasoline.

propane fumes from asphyxiating the driver should the tank be overfilled or develop a leak.

MAKING A DECISION

By now you've had a chance to consider the pro's and con's of propane. You should be at a point of deciding whether or not propane can save you money based on the initial conversion costs, your vehicle's fuel economy, and your annual mileage. You should also be ready to decide if the advantages outweigh the disadvantages for your situation.

If your decision is to go ahead, the following chapters will tell you exactly what is needed to convert to propane, the different types of conversion, tuning, troubleshooting tips, and even some insights into the propane conversion business. There's also a chapter on compressed natural gas for those who want to consider that alternative.

If you've decided against propane, don't be too quick to rush to the "easier" alternative—diesel power (Fig. 4-3). Most diesel engine options carry a healthy cost penalty for the privilege of enjoying higher fuel efficiency. As with propane conversion, the initial added cost of $1000 to $1500 or more must be offset in fuel savings to make the investment worthwhile. Again, it takes a lot of miles to reach the break-even point.

Diesels also have their share of drawbacks. Performance is usually sluggish, they are noisy and smelly, they smoke, they require more frequent oil changes than gasoline engines, and they're costly to fix when something goes wrong. A diesel injection pump, for instance, can cost hundreds of dollars to replace if its ruined by a batch of contaminated fuel. Getting fuel can still be a problem for motorists in some areas.

As for the future, there's no guarantee that diesel fuel will be any more available than gasoline. Some industry experts are predicting a diesel-fuel crunch as a result of the increased demand for the fuel. Diesel fuel may become more costly to use than gasoline if

Fig. 4-2. Propane tanks can eat up a lot of space in passenger car trunks, but they usually create no space problems in trucks or pickups.

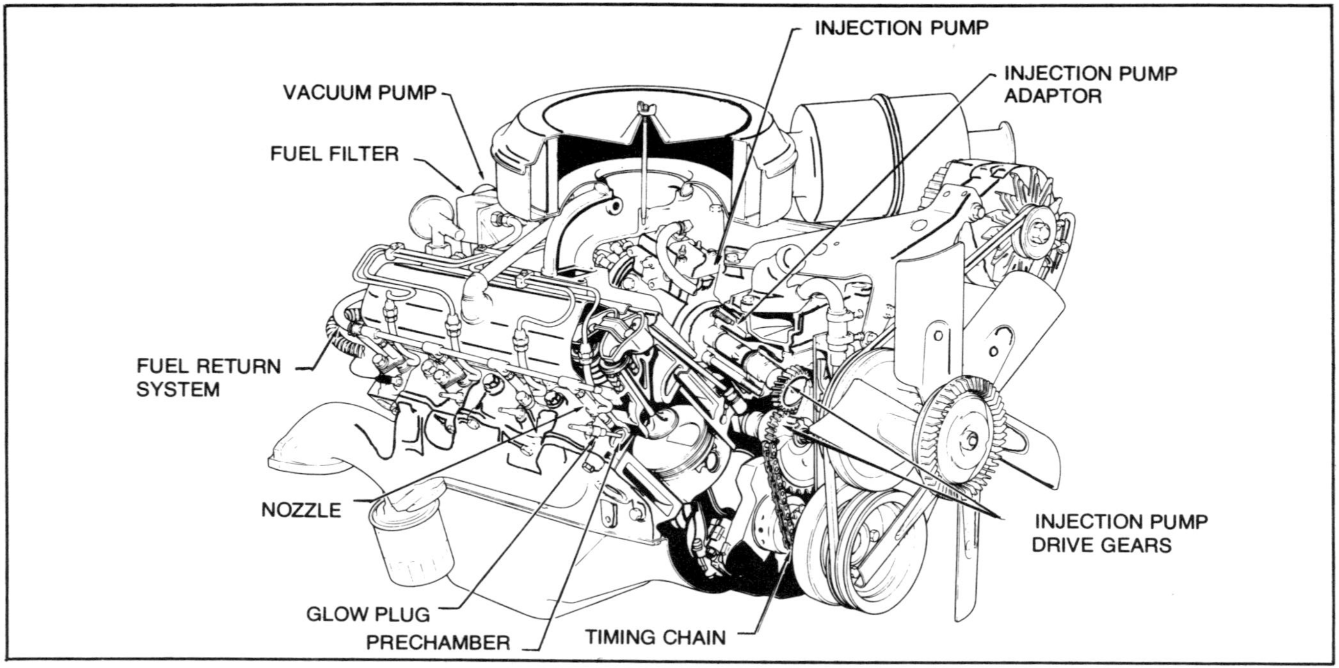

Fig. 4-3. Diesels are sometimes considered to be a more practical alternative to high gasoline prices than propane. But diesels, such as this Oldsmobile 350, are expensive, they have complicated fuel injection systems, and they are costly to repair.

the price continues to rise. In late 1981, the price of diesel fuel had already surpassed regular gasoline in some areas of the country. If the price increases to where it costs 20 to 25 percent more than gasoline, diesel's economic advantage will be lost. And you can't convert a diesel engine to run on straight propane.

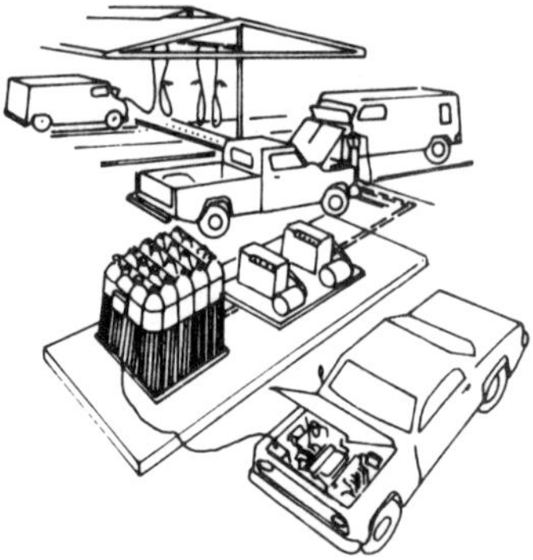

Chapter 5

Propane Conversion Basics

In this chapter, I will explain the basic components and modifications necessary to run an engine on propane. Propane is a gaseous fuel that can be stored as a liquid if it is kept under pressure. The amount of pressure required depends on the temperature. The higher the temperature the greater the pressure required to keep the fuel from evaporating.

In a typical propane fuel system, the operating pressures might range from 75 psi to 240 psi. Because this is far too great a pressure for the thin metal construction of an ordinary gasoline fuel tank to withstand, converting to propane requires the use of a much stronger fuel tank. High pressure fuel lines are also necessary.

Some means must also be provided to convert the high-pressure liquid propane back into a low-pressure gas. Because there's no such device in a standard gasoline fuel system, one must be added. In addition, propane vapor won't work in a carburetor designed for gasoline. Therefore, a special gaseous carburetor must be installed.

Because of such factors, all the components of a gasoline fuel system must be replaced when converting an engine to run on propane. With a dual-fuel setup where the engine will be run on either gasoline or propane, the propane components are added as an auxiliary fuel system.

BASIC FUEL SYSTEM COMPONENTS

There are four basic components in a propane fuel system. All propane fuel systems have these components in common (see

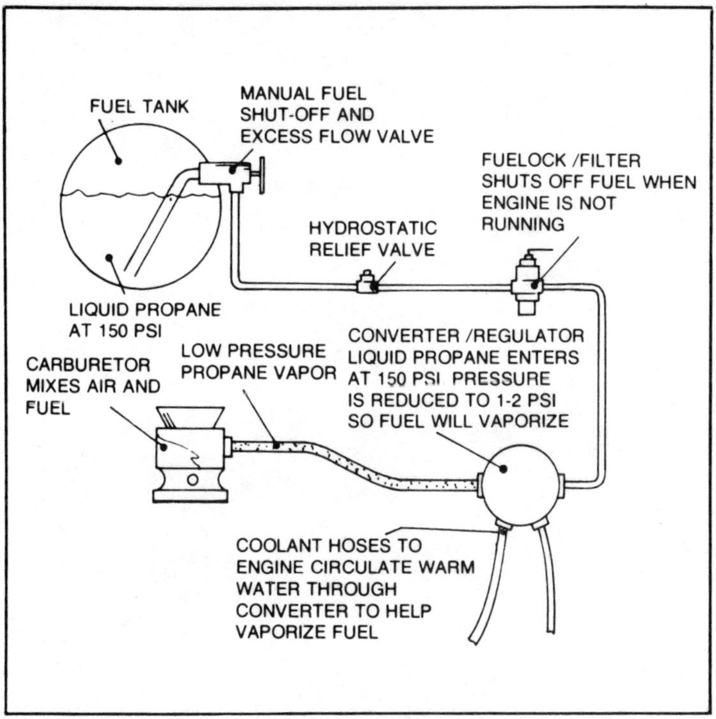

Fig. 5-1. Basic propane fuel system components.

Fig. 5-1). Nevertheless, the design and appearance of the various parts can vary considerably from one equipment manufacturer to another. See Figs. 5-2 through 5-8.

Propane Tank. A heavy steel fuel tank capable of withstanding working pressures of up to 250 psi or more with various safety features to prevent excessive pressure buildup within the tank.

Fuelock. An electric or vacuum-operated fuel shut-off device that stops the flow of propane from the fuel tank line when the engine is not running. A filter can be incorporated into the fuelock to remove inpurities from the fuel.

Converter-Regulator (also called *vaporizer*). A two-stage regulator and heat exchanger that allows the high-pressure liquid propane from the fuel tank line to expand into a low-pressure gas. It is positioned between the fuelock and the carburetor.

Carburetor (also called mixer on dual fuel *applications*). The carburetor's job is to meter the low-pressure gaseous propane to the engine as required. Airflow through the carburetor works against a spring-loaded diaphragm to control fuel flow.

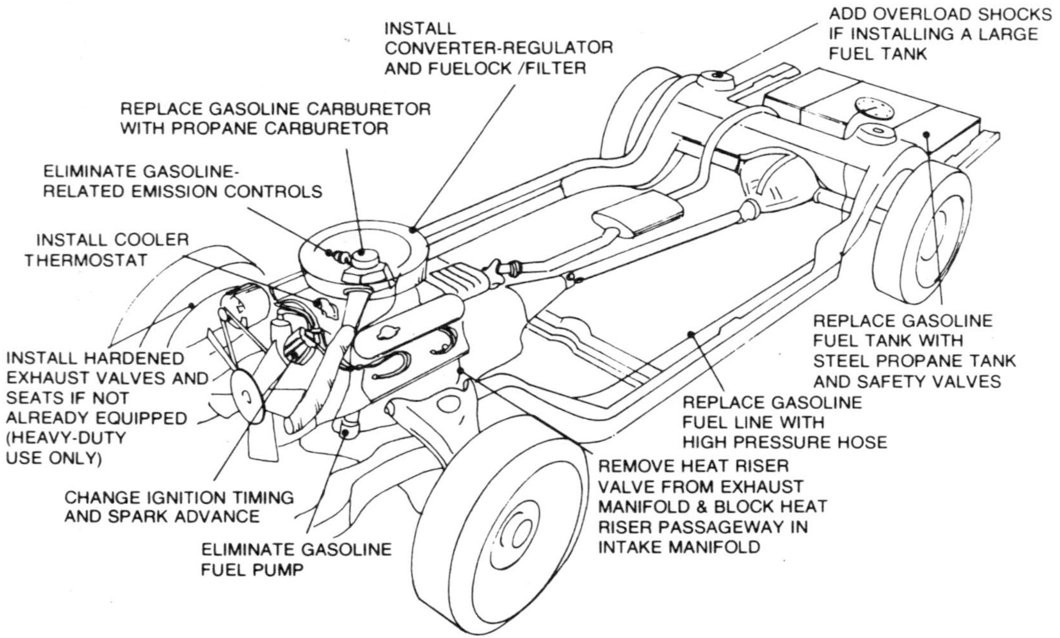

Fig. 5-2. Basic modifications for converting to propane.

Fig. 5-3. A floor mount LP tank for a pickup truck. Shown is a 20-x-65-inch, 82.7-gallon Manchester tank.

LP FUEL TANK

Propane fuel tanks are available in a huge variety of sizes and shapes; they range in capacity from 20 gallons to as much as 200 gallons or more. In most conversions, a tank size somewhere between 25 and 100 gallons is used to give the vehicle a satisfactory driving range. Larger tanks are used on larger vehicles or where an extended driving range is sought. The maximum legal capacity for a passenger-carrying vehicle tank is 200 gallons, with no more than two tanks per vehicle. The maximum legal fuel tank size for any other vehicle is 300 gallons.

Fig. 5-4. Single or double trunk-mounted LP tanks are available for passenger cars. Single tanks range from 20 to 30 gallons in capacity while doubles range from 30 to 40 gallons.

Fig. 5-5. A propane-powered forklift carrying a double propane tank for a car.

The tanks are constructed of heavy gauge steel or aluminum to meet rigid safety standards set by the Department of Transportation (DOT) or the American Society of Mechanical Engineers (ASME). Unapproved tanks must never be used (see Fig. 5-9).

Fig. 5-6. LP tanks are available to fit into the space formerly occupied by the spare tire well. In this installation, the bottom of the well has been cut out to make room for the tank. The tank is installed from above with all valves and lines located outside the trunk compartment.

Fig. 5-7. A pair of LP tanks securely bolted to the frame of a truck. For safety sake, tanks should always be attached with heavy mounting brackets, high-grade steel bolts, and backing plates.

Fig. 5-8. Inside view of an LP fuel tank. The small object that extends into the tank from the side is the float for the fuel gauge. The black pipe that runs up at a slight angle inside the tank is the liquid fuel line. Vapor pressure inside the tank pushes the liquid fuel up the pipe and through the fuel line.

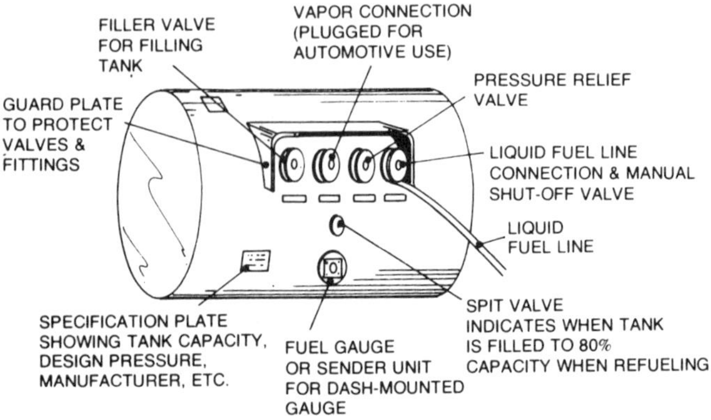

Fig. 5-9. Major components of an LP fuel tank. All approved ASME tanks must have labeled valve connections, a specification plate, and a guard plate to protect the valve connections.

Although the actual working pressures inside a fuel tank are usually less than 250 psi, the tanks are built to withstand pressures four or five times greater. A safety factor of this magnitude in approved tanks makes them practically indestructible. But there is a weight penalty. Most tanks are fairly heavy, and this is especially true for the larger ones. Fortunately, the heavy weight of the tanks is offset somewhat because propane weighs about 2 pounds a gallon less than gasoline. For fuel tanks of equivalent size, a propane tank might actually weigh less than a filled gasoline tank.

For example, a 25-gallon propane tank would weigh approximately 80 pounds when empty and approximately 160 pounds when filled to 80 percent capacity with propane. A 25-gallon gasoline tank, on the other hand, weighs approximately 30 pounds when empty, but it would tip the scales at over 180 pounds when filled with fuel.

Propane tank capacities are rated according to how much water the tank will hold when completely filled. In practice, however, the fuel tanks are never filled to capacity. They are filled to only 80 to 85 percent of rated capacity to allow for expansion (see Fig. 5-10). Such a safety margin is necessary because propane expands as the temperature rises. When an LP tank is filled on a cool morning—for instance, to 80 percent capacity—the fuel could expand to 90 or 95 percent of the tank's capacity by the middle of the afternoon on a warm day.

If no room for expansion were provided within the tank, pressure would rapidly increase. Eventually something would have to give. That something would be the safety release valve, or pop-off valve, that is installed on every approved LP fuel tank. Releasing the excess propane from the tank prevents a dangerous buildup of pressure, but it also wastes some of the fuel. When propane is vented under such circumstances, it can also create a potential hazard for anyone standing nearby—especially if they're smoking. And don't forget that liquid propane can cause frostbite if it is sprayed on bare skin.

Every approved LP fuel tank has a number of safety features built into it to prevent both excessive pressure buildup inside the tank as well as the loss of fuel in the event of an accident. In fact, the DOT says approved LP fuel tanks are probably three times safer than gasoline tanks.

Every tank has a safety release valve that is designed to vent excess internal pressure to the atmosphere (see Fig. 5-11). The safety is a simple spring-loaded valve that is preset to open at a specified pressure. For fuel tanks mounted inside automobiles or trucks, the valve opens if tank pressure exceeds 312 psi. For externally-mounted tanks, the valve opens if the pressure exceeds 250 psi. The valve opens only long enough to allow the excess pressure to escape. Once the pressure drops below the preset limit, the valve closes and reseals the tank.

Overfilling the fuel tank or subjecting it to high temperatures (as when the vehicle sits in the hot sun on a summer day) can cause the safety release valve to vent some propane. To minimize such

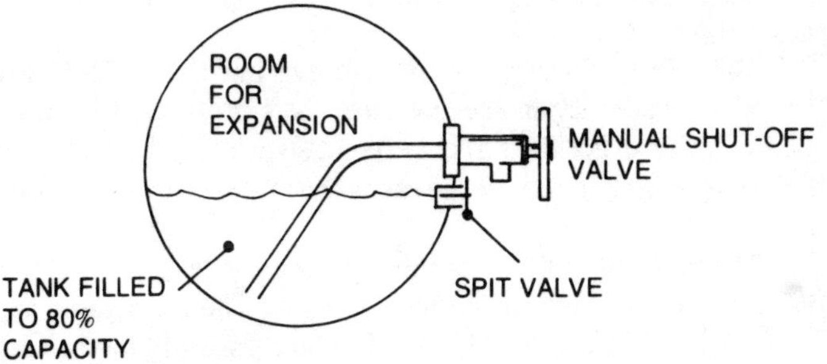

Fig. 5-10. LP fuel tanks should never be filled beyond 80 to 85 percent capacity to allow room for expansion. The spit valve will puff liquid fuel when the tank is filled to the correct level.

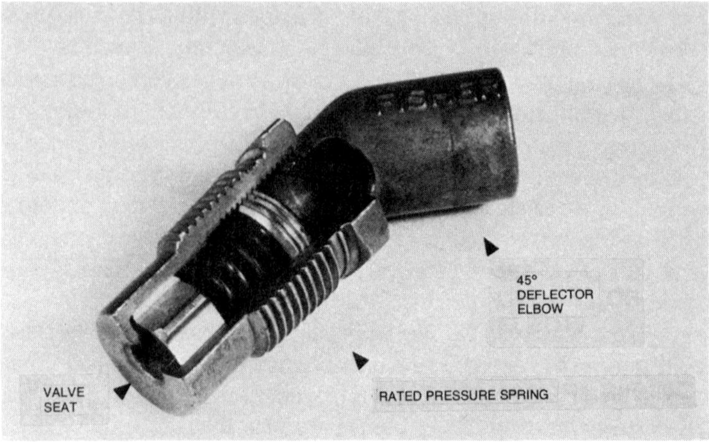

Fig. 5-11. The safety relief valve vents propane from the fuel tank if internal pressure exceeds 250 psi for externally mounted tanks or 312 psi for internally-mounted tanks. The elbow directs fuel away from the vehicle's body or adjacent vehicles.

problems, most fuel tanks are painted white or silver to reflect some of the sun's heat.

The safety release valve is sometimes equipped with a deflector elbow to direct any escaping propane away from the vehicle or the ground. A 45-degree elbow fitting is commonly used. In vehicles such as buses where people might be standing nearby, a vent line is routed from the safety valve to a point above the roof line of the vehicle. In automobiles or pickup trucks with toppers where the fuel tank is mounted inside the vehicle, the vent line is routed to the outside so that any propane fumes that escape cannot accumulate inside the vehicle.

The safety release valve itself screws into the fuel tank so that the valve is actually inside the tank. This protects it from being sheared off in an accident. If it were not so protected, a hard blow to the valve might knock it off and cause the propane to spray out of the tank.

The fuel tank is filled through a *filler valve* (Fig. 5-12) that, like the safety valve, is threaded into the tank. The filler valve has an internal double backflow check valve that prevents the accidental loss of propane. In other words, it allows the fuel to flow only one way: into the tank and not out. On cars and pickup trucks with toppers where the fuel tanks are mounted inside the vehicle, a *remote filler* location is used to simplify refueling and to prevent any

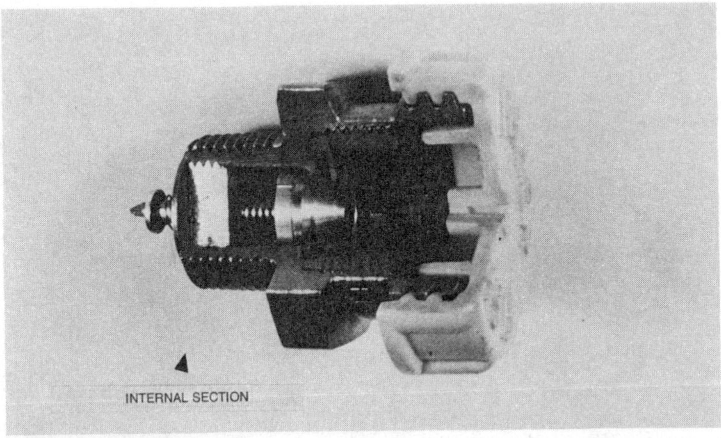

Fig. 5-12. The filler valve has an internal, double back-flow check valve that prevents the accidental loss of fuel.

fuel spills inside the vehicle (Fig. 5-13). A remote filler is nothing more than a filler connection mounted on a fender or under the bumper. The remote fill is connected to the valve on the tank with a length of hose.

LP fuel tanks are also equipped with a *bleeder spit valve*. The purpose of this valve is to signal when the tank is filled to 80 or 85

Fig. 5-13. A remote filler location is used for tanks mounted inside a vehicle. This eliminates the possibility of propane fumes getting inside the vehicle during refueling.

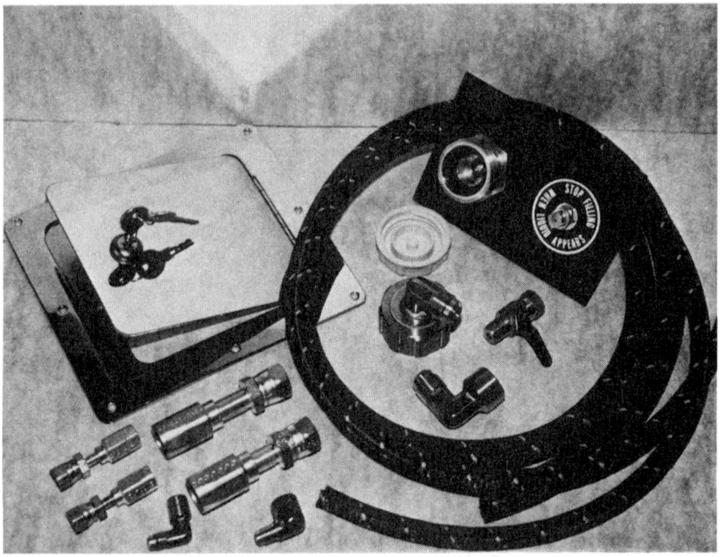

Fig. 5-14. A typical remote filler kit includes the filler valve, spit valve, hoses and connections, and a locking cover.

percent capacity by spitting out a small amount of liquid propane (Fig. 5-15). When the person filling the tank sees liquid come out of the spit valve, it's time to turn off fuel flow. Most refueling stations are also equipped with an automatic shut-off that stops the flow of fuel when the tank reaches a preset pressure.

The fuel line that runs from the LP tank to the engine compartment is connected to a manual shut-off valve at the tank (Fig. 5-16). This valve can be used to shut off the fuel line when the tank is not being used. The valve also has an internal *excess flow check valve* built into it that prevents fuel from gushing out of the tank in case the fuel line ruptures or is severed in an accident.

As with gasoline fuel tanks, LP tanks can be fitted with various types of fuel gauges. An internal float is used to read the fuel level, and is connected either to an external gauge or sending unit for a dash-mounted gauge (see Figs. 5-8 and 5-17). In dual fuel conversions, the existing gasoline fuel gauge can be modified to read either gasoline or propane by adding a switch and the necessary electrical resistor.

All the various tank fittings and valves are protected by a heavy steel plate that is welded or bolted to the tank. The fuel tank is mounted to the vehicle's frame or chassis with 7/16-inch or larger high-tensile-strength bolts and backing plates.

When a propane tank is mounted inside the trunk of an automobile, the National Fire Protection Association (NFPA) standards state the trunk must be sealed off from the passenger compartment by a vapor barrier. Almost any sealing material can be used to make the vapor-tight seal, but the most common and easily applied is probably urethane foam (Fig. 5-18). The vapor barrier prevents any fumes, that might accidentally leak from the tank, from getting inside the passenger compartment. A small vent hole in the trunk would allow any propane fumes to escape to the outside.

For vans, jeeps, station wagons, or cars with hatchbacks, a vapor-tight box of plywood, aluminum, or steel must be constructed around the tank to seal it off from the rest of the vehicle (Fig. 5-19).

Fig. 5-15. The bleeder spit valve signals when the tank is filled to 80 or 85 percent capacity by puffing out a small amount of liquid propane. The valve is twisted open by hand when refueling starts. It is closed once refueling is complete. The valve opening must not be larger than a #54 drill bit.

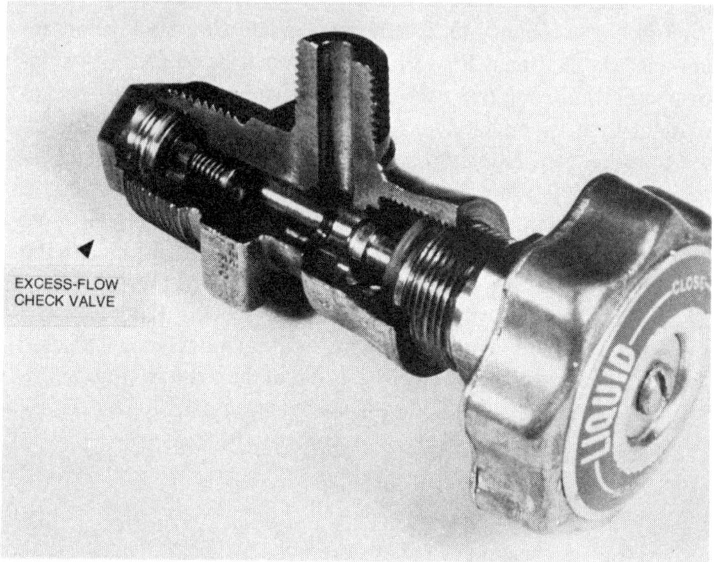

Fig. 5-16. Manual shut-off valve for the fuel line with its internal excess flow-check valve.

The compartment is then vented to the outside so fumes would be able to escape.

All of these safety devices and standards that apply to LP fuel tanks might sound like a lot of trouble, but they are based on

Fig. 5-17. An LP tank with a tank-mounted fuel gauge. Note the protective guard plate around the tank connections.

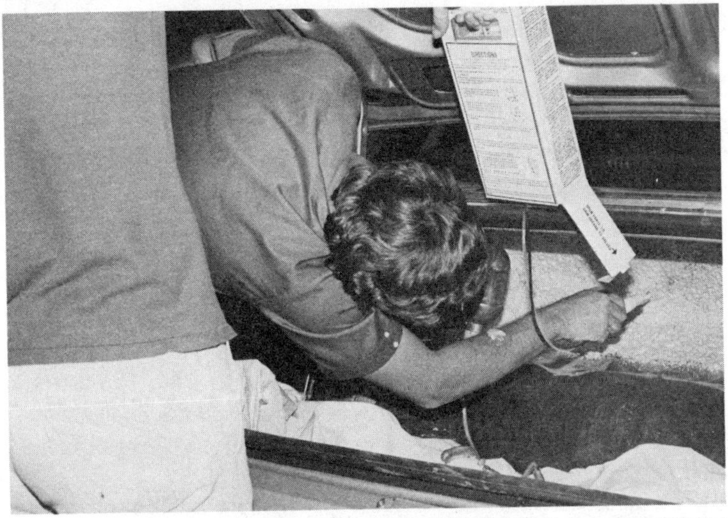

Fig. 5-18. Forming a vapor barrier to seal off the trunk in a passenger car conversion.

procedures designed to eliminate needless hazards. If a vehicle is to be converted to propane, it should be done in accordance with the accepted standards and regulations outlined in the NFPA booklet #58. Any job worth doing is worth doing right.

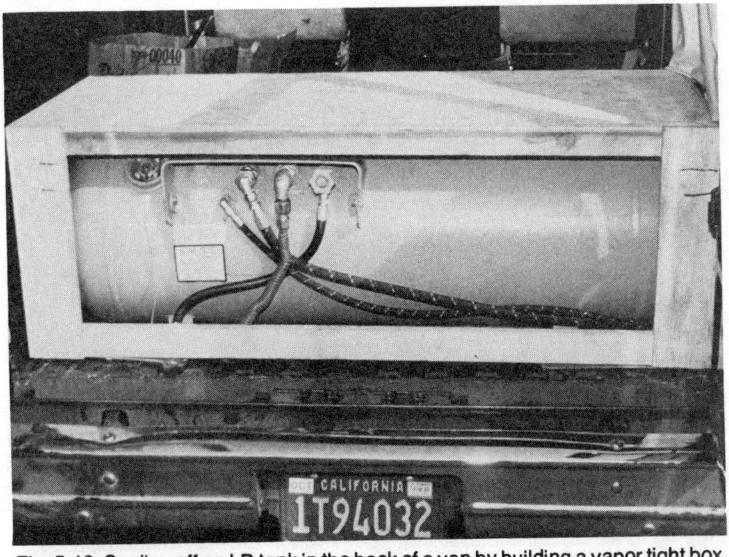

Fig. 5-19. Sealing off an LP tank in the back of a van by building a vapor tight box around the tank.

FUEL LINE

As with fuel tanks, there are certain standards that apply to LP fuel lines. For one, they must be capable of withstanding high pressures. The NFPA standards require lines to be capable of handling working pressures to 350 psi with a burst pressure rating of 1750 psi. In Texas, the law requires a working pressure of 500 psi and a 1250-psi burst strength.

LP fuel lines are typically constructed of steel reinforced braid with screw-on pressure fittings on the ends (Fig. 5-20). Push-on type fittings should never be used with propane fuel lines because the high pressures could cause the hose to pop off.

All fuel line fittings are sealed with pipe dope to make a gas-tight seal. Where the fuel line passes through the body sheet-metal, reinforced fittings, brackets, or protective coatings can be used.

Because the fuel line carries high-pressure liquid propane at the same pressure as the fuel tank, a hydrostatic pressure relief valve is usually installed somewhere along the length of the fuel line to vent excess pressure if it develops. The safety release valve is set to open between 400 and 500 psi. Pressure can build up in the fuel line if the tank valve is closed because there's no room for expansion within the line itself.

FUELOCK

The fuel line leads to a fuel shut-off valve called the fuelock. The fuelock is typically located in the engine compartment and is either vacuum operated or electrically operated (Fig. 5-21). The

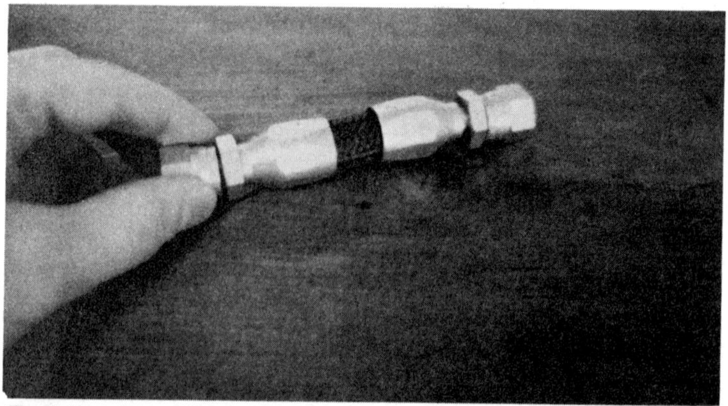

Fig. 5-20. A piece of braded fuel hose with screw-on, aircraft-type fittings. Never use push-on fittings with high-pressure LP fuel lines.

Fig. 5-21. LP fuelocks. The two fuelocks at the left use electric solenoids while the one at the right has a vacuum diaphragm. The fuelock shuts off the flow of fuel from the tank when the engine is not running.

fuelock shuts off the flow of fuel from the tank line when the engine is off.

On vacuum-operated fuelocks, the flow of fuel is blocked when there is no engine vacuum whether the ignition switch is on or off. This means fuel cannot flow to the converter until the engine is cranked, and it will stop the fuel flow if the engine dies.

On electrically operated models, the valve closes when the ignition is turned off. A vacuum or oil pressure switch should also be used to make sure the fuelock shuts off when the engine is not running.

A fuelock can have a small filter incorporated into it to screen out impurities that reach the fuel tank. Such filters usually go 100,000 miles or more between replacements (Figs. 5-22 and 5-23). Use a fuelock that has been approved by Underwriters Laboratories. Unapproved fuel shut-off devices should never be used.

On dual-fuel conversions, an additional fuel shut-off solenoid is required on the gas fuel line to block the flow of fuel when switching from gasoline to propane. The solenoid is controlled electrically from a dash-mounted switch.

CONVERTER-REGULATOR

The converter-regulator's job is to change the high-pressure liquid propane into a low-pressure gas before it enters the carburetor or mixer (Figs. 5-24 through 5-28). It does this by reducing the fuel pressure as it enters the converter so that the propane can vaporize inside a small chamber. As the liquid propane changes into

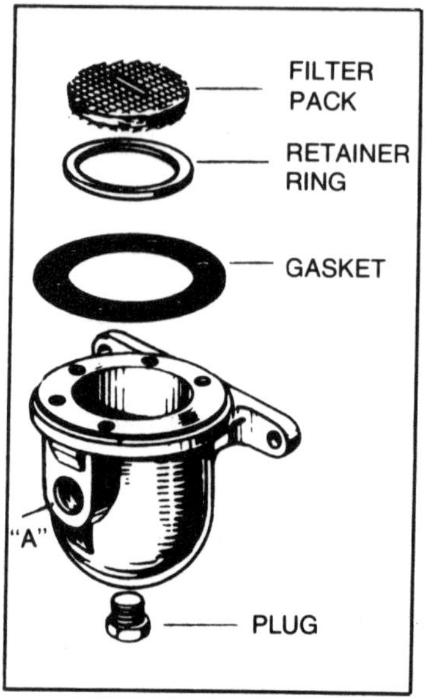

Fig. 5-22. Typical fuelock filter assembly. Filter life is usually 100,000 miles.

a gas, it absorbs heat. This creates a cooling effect inside the converter. To counteract this cooling effect and to prevent a possible freeze up of the converter, hot water from the engine's cooling system is circulated through the unit. A couple of lines are tapped into the engine's heater hoses so that some of the coolant will circulate through the converter. Inside the converter, heat exchanger fins transfer the warmth from the coolant to the fuel. At no time does the coolant make direct contact with the fuel (if it does, there's probably a bad gasket in the converter).

Installation of the converter should always be in accordance with the manufacturer's instructions. The converter can be mounted on the firewall, on the fenderwell, or on top of the engine. Mounting it in the proper orientation (right side up) is important because fuel can pool inside the converter. In some cases this can cause problems.

When connecting the hot water lines, brass fittings rather than steel should be used to prevent electrolysis corrosion of the converter's aluminum housing. The hot water lines should be tapped into the heater hoses so that coolant will flow through the converter whether or not the heater is in use.

Once the fuel has been vaporized inside the converter, it is metered to the carburetor or mixer on demand. A large hose carries the fuel vapor from the converter to the carburetor. A mechanical or

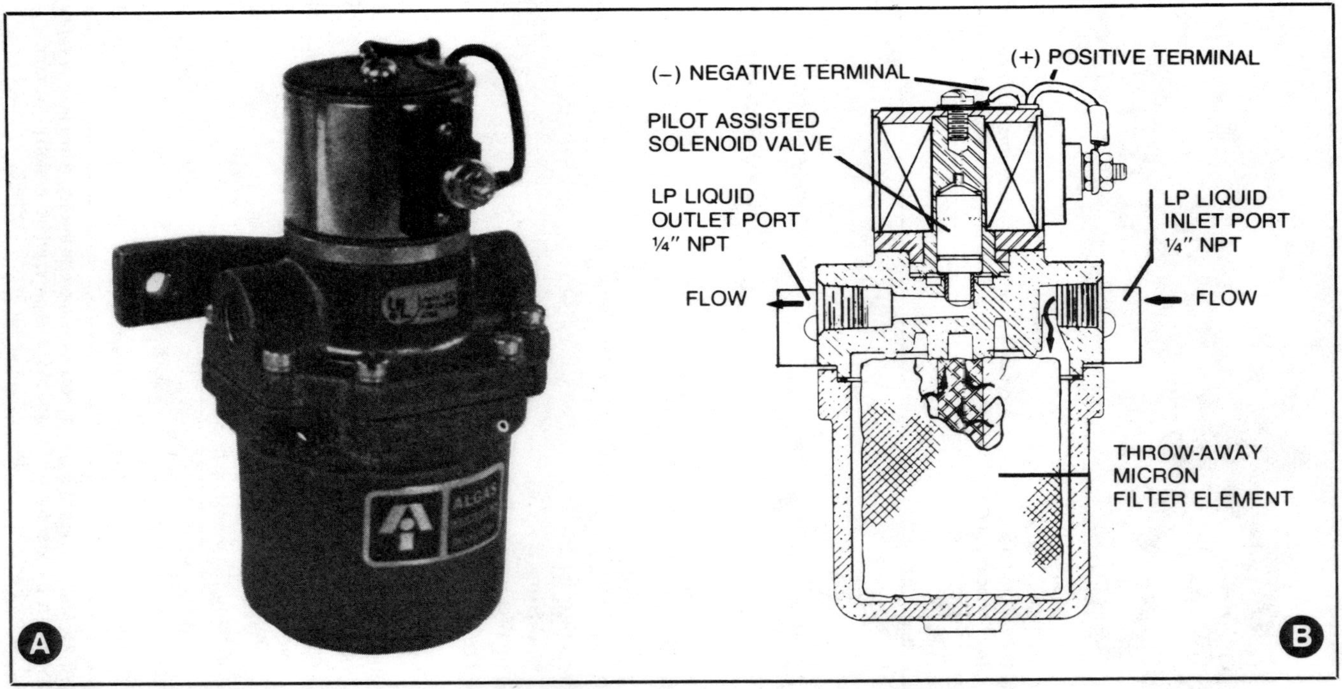

Fig. 5-23. This fuelock from Algas (A) has a replaceable filter element (B) that resembles a miniature oil filter.

Fig. 5-24. Converter-regulators. From left to right are Century, Impco, Beam, and another Impco model. The converter changes the high-pressure liquid propane into a low-pressure gas.

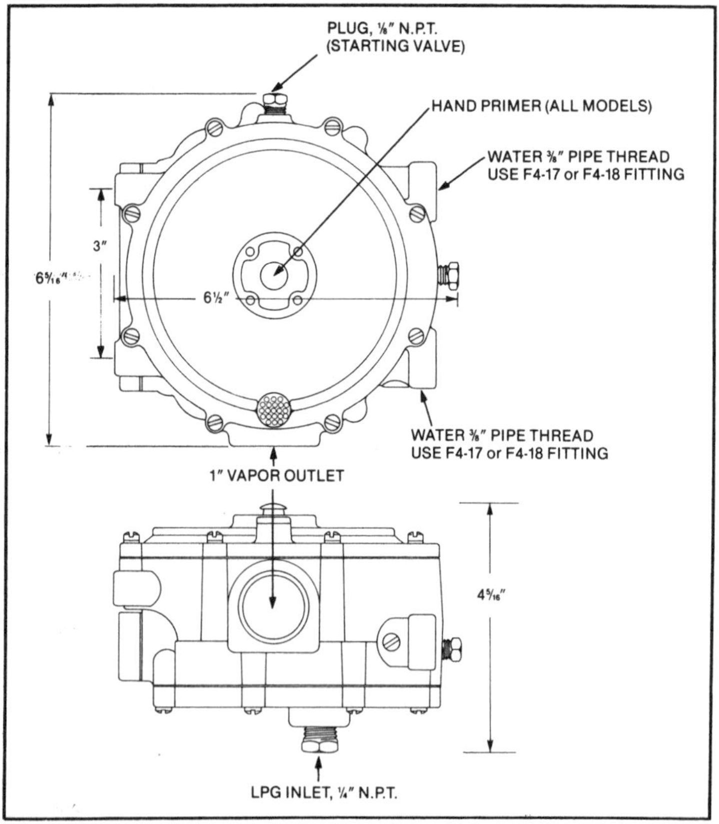

Fig. 5-25. Main fittings of a typical converter include the liquid fuel inlet, a vapor outlet, and two water-hose connections. The hand primer button in the center is used for cold-weather starting. Shown is an Impco Model L.

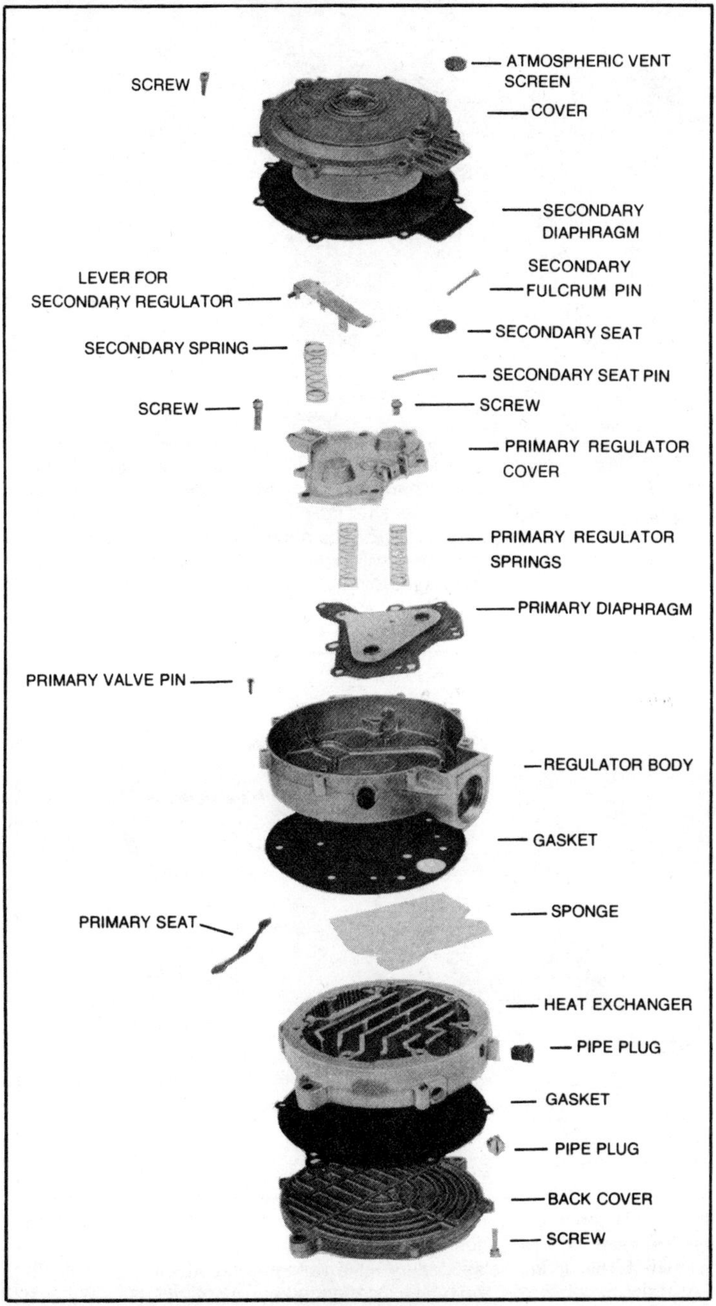

Fig. 5-26. An exploded view of a typical converter. Shown is an Impco Model E.

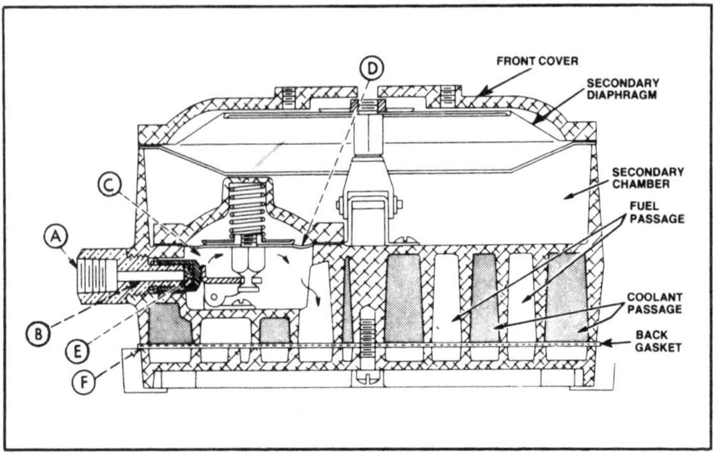

Fig. 5-27. High-pressure liquid propane enters the converter through the inlet at A and expands into the heat exchanger body between the coolant passages. The pressure of the expanding fuel pushes the primary diaphragm D up against a spring and closes the primary valve E. This happens when the fuel pressure reaches about 5 to 8 pounds. Valve E will remain closed (shutting off fuel into the converter) until the carburetor begins to draw vaporized fuel from the secondary chamber.

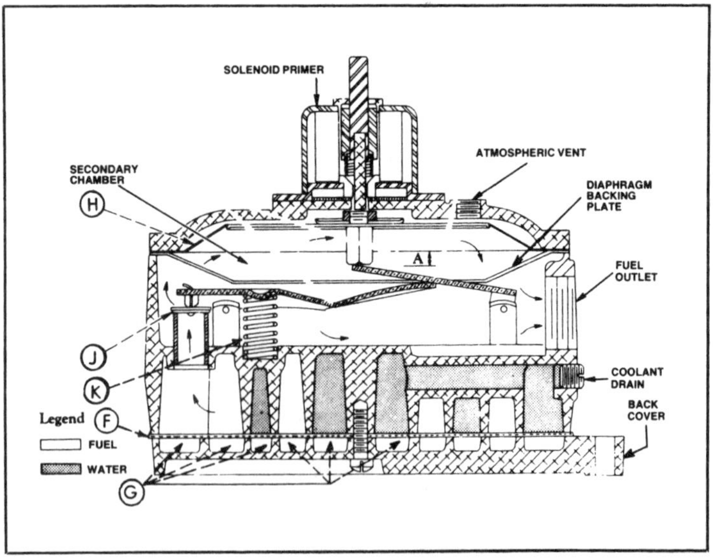

Fig. 5-28. When engine vacuum begins to draw vaporized fuel from the secondary chamber, the reduction in pressure pulls the secondary diaphragm H downward. This opens the secondary valve J allowing vaporized fuel to enter the secondary chamber from the heat exchanger body. The primer (solenoid operated in this case) manually opens the secondary valve J to start the fuel flowing.

electrical primer on the converter supplies an extra shot of fuel for cold-weather starts.

LP CARBURETOR

With a straight propane conversion, the LP carburetor replaces the gasoline carburetor (Figs. 5-29 and 5-30). To make the conversion, the gasoline carburetor is removed from the engine. The lower throttle plate is then disassembled from the carburetor body. The LP carburetor is bolted to the throttle plate and the whole assembly is then reinstalled on the engine.

In a dual-fuel conversion, the LP mixer is mounted inside the air cleaner over the existing gasoline carburetor (Figs. 5-31 and 5-32). Or in cases where space is limited, it is mounted between the air cleaner and carburetor with an offset adapter. In other words, the carburetors are stacked so that incoming air must pass through both before entering the engine. Regardless of the type of conversion, the carburetor's job is still to mix the right amounts of propane and air so that it can be fed to the engine as needed.

Fig. 5-29. A typical LP carburetor. The carburetor body fits onto the throttle plate of the gasoline carburetor or it can be ordered with its own plate assembly. Shown is an Impco Model CA425 four-barrel carburetor.

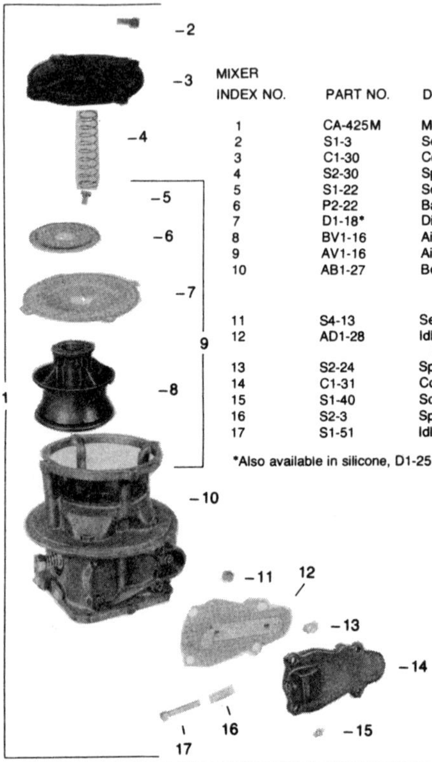

MIXER INDEX NO.	PART NO.	DESCRIPTION
1	CA-425M	Mixer complete
2	S1-3	Screw, 10-24 × 5/8" SEMS (4)
3	C1-30	Cover, air valve diaphragm
4	S2-30	Spring, air valve
5	S1-22	Screw, 6-32 × 1/4" SEMS (4)
6	P2-22	Backup plate
7	D1-18*	Diaphragm, air valve
8	BV1-16	Air-gas valve: V1-16, V2-19, S4-14
9	AV1-16	Air-gas valve assy., with diaphragm
10	AB1-27	Body assy.: B1-27, C1-29, J1-5, S1-50, S2-16, W1-33, S3-27, V2-18, P3-3(3). P3-10, S1-69(3)
11	S4-13	Seat, idle valve
12	AD1-28	Idle diaphragm assy.: D1-28, L1-30, P2-23, R2-4(2), S4-13
13	S2-24	Spring, idle valve
14	C1-31	Cover, idle valve
15	S1-40	Screw, 10-24 × 3/8" SEMS(4)
16	S2-3	Spring, idle adjustment
17	S1-51	Idle screw

*Also available in silicone, D1-25

Fig. 5-30. An exploded view of an Impco Model CA425 air valve LP carburetor.

The typical LP carburetor or mixer has two basic adjustments: an idle adjustment and a power adjustment. The idle screw is used to obtain the desired air/fuel ratio when the engine is idling. The power adjustment is used to set the air/fuel ratio for part and full throttle operation.

The typical propane carburetor has a spring-loaded neopreme diaphragm to sense air flow through the unit (Fig. 5-31). As the air flow increases, the diaphragm is pushed up. This lifts a gas metering valve off its seat and allows more fuel to enter the carburetor. The resulting pressure drop in the fuel hose between the carburetor and converter causes the converter to release more fuel through to the carb. As more air passes through the carburetor, more fuel is metered to the carburetor so the proper air/fuel ratio can be maintained. It might sound like a complicated process, but actually it's quite simple.

Compared to a gasoline carburetor with its complex metering circuits, main jets, idle jets, power enrichment valves, automatic choke, choke pulloff, needles and floats, the propane carburetor is a shining example of simplicity. The LP carburetor's only moving parts are the air-sensing diaphragm and gas metering valve. Service, in most instances, is limited to replacing a worn diaphragm. This is a simple job that takes less than 15 minutes and it can be done without removing the carburetor from the engine. New silicone-rubber diaphragms last the life of the vehicle in most cases.

Through the use of various adapter plates, a relatively small number of basic LP carburetors can be made to fit practically any engine application. This can greatly simplify the stocking of replacement parts and simplify service for large fleets.

In some applications, such as where the engine is turbocharged or where there is a high degree of restriction in the air cleaner, a *balance line* must be installed between the LP carburetor and the atmospheric vent on the converter to maintain the correct fuel

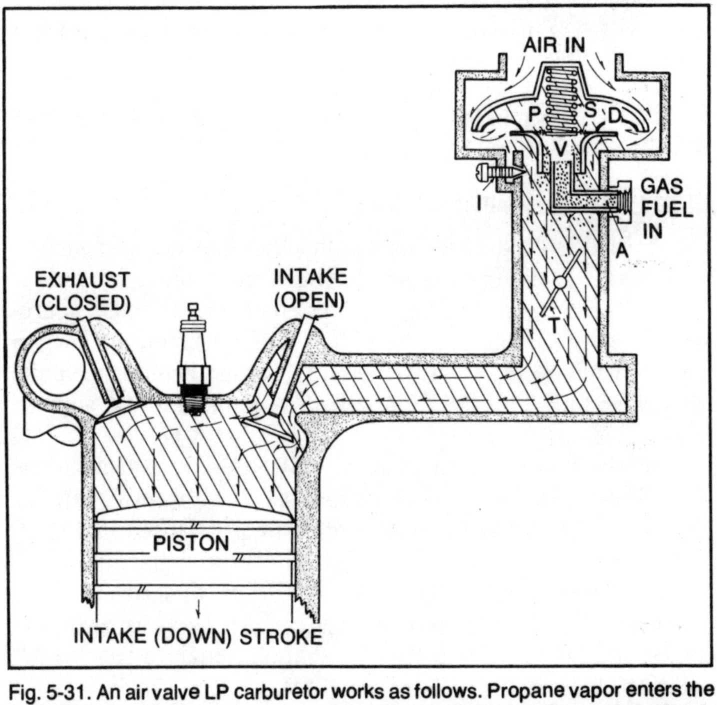

Fig. 5-31. An air valve LP carburetor works as follows. Propane vapor enters the carburetor through the gas fuel inlet A. Air flowing into the engine causes the air valve diaphragm D to rise. This lifts the gas metering valve V off its seat so that fuel can flow in and mix with the air.

Fig. 5-32. A typical LP fuel mixer for a dual-fuel conversion. Shown is an Impco Model CA300 without its air cleaner.

metering. If a vacuum-operated fuelock is used, the balance line also must be routed to the fuelock vent.

IGNITION SYSTEM MODIFICATIONS

In addition to the fuel system modifications and components just described, certain other modifications are recommended to get the most performance out of a converted engine. Because of propane's slower burning characteristics (higher octane), the ignition timing should be modified. Generally, propane likes more initial advance and a faster spark advance than gasoline. Total advance should be about the same.

In heavy-duty engines that typically operate at speeds less than 3800 rpm, the initial spark advance should be about 10 degrees BTDC (before top dead center), with a total advance of 27 to 30 degrees by 2000 or 2200 rpm. In a passenger car engine, the initial advance would be about 10 degrees BTDC with a total advance of 38 to 40 degrees by 2800 to 3400 rpm. The actual timing settings would depend on the application because different engines require different adjustments (Figs. 5-33 and 5-34).

The initial timing advance is adjusted by simply loosening the distributor hold-down nut and retiming the engine with a timing

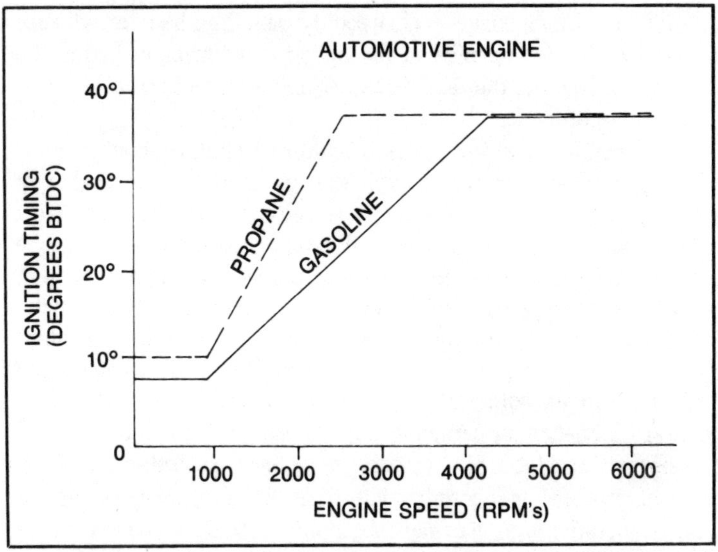

Fig. 5-33. Comparison of spark advance curves for an automotive application. When converting to propane, the distributor should be recalibrated for faster spark advance, and the engine retimed for more initial advance. Total advance, however, remains about the same.

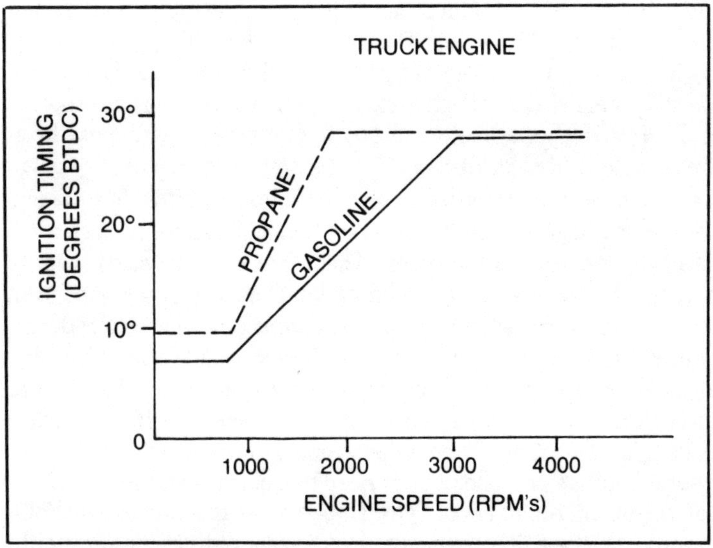

Fig. 5-34. Spark advance curves for a truck engine. Because the engine's operating speed is much lower, total advance is less than in automotive applications. But propane still requires more initial lead timing and faster advance to perform well. Total advance remains about the same.

light. The advance curve is changed by installing heavier advance weights or lighter springs in the advance mechanism inside the distributor. Ideally, the distributor would be recalibrated on a distributor machine.

Although the stock ignition system (whether conventional point ignition or electronic high energy ignition) is capable of igniting the fuel, it should be inspected closely to make sure that it is performing up to specifications. Pitted or worn contact points, a cracked distributor cap, dirty or worn plugs, a weak coil, or high resistance in the ignition cables can cause problems with any fuel.

Most experienced mechanics recommend installing "colder" spark plugs when converting to propane. For heavy-duty trucks, use two heat ranges colder. For pickup trucks, use one heat range colder. For passenger cars, use one heat range colder if the car will be subjected to a lot of highway driving or hard use. Otherwise stick with the stock plug recommendation. In dual fuel conversions, no ignition modifications are made if gasoline is to be used frequently.

INTAKE SYSTEM MODIFICATIONS

The only modifications recommended for the intake system are those that help keep temperatures down. The heat riser plumbing that feeds hot air from around the exhaust manifold into the air cleaner can be eliminated. The heat riser plumbing is necessary with gasoline to prevent stumbling and hesitation during the warm-up period and to control exhaust emissions. Because propane is already a vapor when it reaches the carburetor, the extra heat isn't needed (Fig. 5-35).

On most V-8 and V-6 engines, the intake manifold has a small passageway cast into it that allows hot exhaust gases to pass from one cylinder head to the other. These hot exhaust gases help to preheat the intake manifold and carburetor so that gasoline will vaporize more quickly. This speeds engine warm-up and reduces emissions while the engine is cold. A heat riser valve is usually installed on one of the exhaust manifolds to force the exhaust back through the intake manifold passage. The valve remains shut until the engine warms up. Then it opens and allows the exhaust to flow freely. Because propane doesn't need this extra heat, the heat riser valve can be removed or wired open, and the intake manifold passage blocked. Replacement intake manifold gasket sets with blocked heat riser ports are available for this purpose. Just removing or wiring open the heat riser valve will eliminate most of the intake manifold preheating.

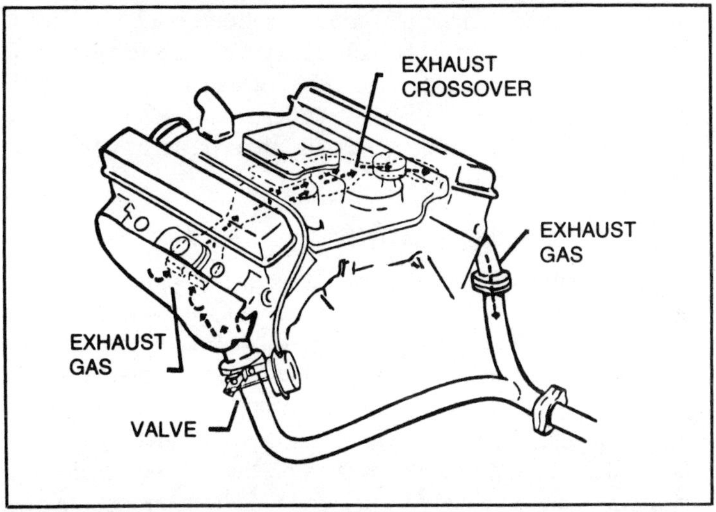

Fig. 5-35. The heat riser valve on the exhaust manifold should be wired open or removed to eliminate intake manifold heating.

In dual fuel conversions, these modifications are not usually made. Therefore, the engine will continue to run satisfactorily on gasoline when it is cold.

COOLING SYSTEM MODIFICATIONS

The only recommended change here for straight propane conversions is to install a colder thermostat. A 160- or 170-degree thermostat should be used in place of the existing unit. Lowering the operating temperature helps the engine run better and resist detonation and preignition. Many late model "emissions engines" come with thermostats as high as 200 degrees because higher temperatures help reduce emissions in gasoline engines. Because propane is a clean-burning fuel, the higher temperature is unnecessary and hinders performance.

EXHAUST SYSTEM MODIFICATIONS

In addition to the removal of the exhaust manifold heat-riser valve, some mechanics prefer to install a low restriction exhaust system to maximize engine performance and reduce exhaust back pressure. A restrictive exhaust system causes the engine to work harder to force the exhaust gases out of the cylinders. This robs power and cuts fuel mileage. By installing larger-diameter exhaust pipes, dual exhausts, or low-restriction mufflers, this lost horsepower is regained for more performance and better mileage. Reducing the back pressure in the exhaust system also minimizes the effects of *valve erosion* in propane-powered engines by allowing the valves to run cooler. Exhaust modifications are not absolutely necessary, but they can be used with either straight or dual-fuel conversions.

INTERNAL ENGINE MODIFICATIONS

For engines designed to burn unleaded gasoline, no internal engine modifications are necessary. The only recommendation is that a new engine be "broken in" on gasoline before it is converted to propane. The reason for doing so is to allow the piston rings to seat completely. Gasoline washes the protective oil coating off the cylinder walls so that friction can wear down the high spots on the rings. With propane the oil remains on the cylinder walls. This greatly prolongs engine life, but reduces friction so that the rings are slow to seat in a new engine. If hard chrome rings are used, the rings might never fully seat. This results in high oil consumption and loss of power. If a new or rebuilt engine must be broken in on propane, regular cast iron rings should be used.

In engines designed to run on leaded gasoline, valve wear can be a problem when propane is used. Besides increasing the octane rating of the fuel, lead acts as a lubricant to slow valve wear. When switching to unleaded gasoline or propane, there's no lead to protect the valves. Valve wear increases unless hardened valves and valve seats are used. Newer engines that are designed to run on unleaded or low-lead fuels already have these hardened parts installed so conversion to propane presents no problems. But in the older engines, accelerated valve wear can be a real problem (see Fig. 4-1).

For short trip, light-duty applications, you can usually get by without the harder valves and seats. For heavy-duty use or high-speed driving, the harder parts should be used. To save money, most such vehicles are driven until the valves begin to show signs of wear (compression will drop). When the heads are removed for servicing, the harder valves and seats are installed. Valve life up to that point will probably be reduced to 30,000 to 50,000 miles.

Keep in mind that the harder valves are only required in older engines designed to burn leaded gasoline. Newer engines that are built to run on unleaded or low-lead fuels should not require modification.

In some small stationary engines, valve rotaters are used on the valve springs to prolong valve life. The rotaters cause the valves to turn slightly with every opening. With propane, however, the added friction caused by the valves turning on their seats can accelerate valve wear. Therefore, removing the rotators is usually recommended.

Internal engine modifications are not required for dual-fuel conversions of older engines if the engine is operated periodically on leaded gasoline. Using a tankfull of leaded gasoline every couple hundred miles will recoat the valves with enough lead deposits to offset the clean-burning effects of propane.

If an engine is being overhauled with the intention of converting it to propane, high-compression pistons or cylinder heads can be installed to take advantage of propane's higher octane rating. Compression can also be increased by milling metal off the cylinder head faces. A static compression ratio of 11 to 1 is probably a safe maximum to aim for. Increasing the engine's compression ratio enables it to burn the fuel more efficiently. This improves both fuel mileage and horsepower.

SUSPENSION MODIFICATIONS

Modifications to a vehicle's suspension are normally not necessary unless the vehicle is equipped with a fairly large propane tank. With pickup and light-duty trucks, the added weight of a propane tank of up to 100 gallons usually helps the vehicle to ride better because of the added weight in the truck bed or box.

For passenger cars, however, added weight can cause the rear end to sag. This is especially true for dual-fuel conversions where the combined weight of a full propane tank and full gasoline tank can add several hundred pounds of extra weight. For such applications, ride height and quality can be restored by installing air-assist rear shock absorbers, overload rear shocks or overload coil springs. Such modifications can be done for under a $100, in most cases, or less if the vehicle's owner installs the parts.

CONVERSION KNOW-HOW

Propane conversions are not something the novice mechanic should attempt. A conversion should only be done by someone who is familiar with the equipment and procedures necessary to do the job right. For the average motorist, it's easiest to have the conversion done at a local conversion center. For fleets, mechanic training is available from a variety of sources including state LP-gas associations and local conversion centers.

Having the right equipment to complete the conversion is just as important as know-how. Adjusting the carburetor or mixer, for example, usually requires the use of an infrared exhaust analyzer to read CO, and a chassis dynamometer to set the power adjustment with the vehicle running at full throttle under load. The trial-and-error approach to engine tuning can be used, but the results might be unsatisfactory.

FACTORY CONVERSIONS

Propane-powered trucks are currently available from Ford and International Harvester (Fig. 5-36), and Ford has recently began selling propane-powered Granadas to fleets in Canada and the United States. The factory-converted cars sell for $600 to $800 over the base price and have a 2.3 liter, 4-cylinder engine with 10.5 to 1 compression. According to Ford, the cars run smoother, accelerate faster, and start easier than their gasoline counterparts.

Ford's propane-powered F-series, LN-series and C-Series trucks are available with either a 370- or a 429-cubic-inch engine. Modifications include an optional 59-gallon LP tank, recalibrated distributor advance curve, 10° F lower engine thermostat, modified intake manifold to eliminate the EGR valve and exhaust crossover passages, nonchrome-plated piston rings, hard-faced alloy valve

Fig. 5-36. An International Harvester truck with factory-installed propane fuel system.

Fig. 5-37. A Holley propane carburetor. This unit uses essentially a gasoline carburetor with modified fuel metering blocks. The carburetor works on the venturi principle (rather than an air valve) with the converter feeding propane to the carb depending on engine vacuum.

seat inserts, elimination of all gasoline-related emission controls, Impco model VFF-30 fuelock, Impco model EOS-20 converter, and an Impco CAG 425-20, 4-barrel carburetor. See Fig. 5-37.

Chapter 6

Straight Versus Dual-Fuel Conversion

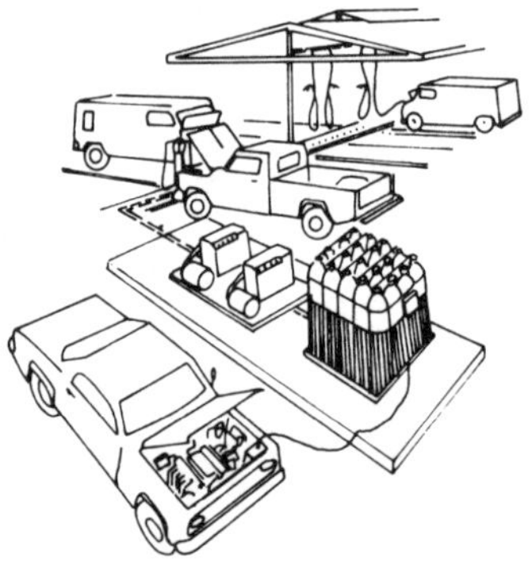

When converting to propane, the alternatives are a straight conversion or a dual-fuel conversion. See Table 6-1. A straight conversion means modifying the vehicle's fuel system so that the engine will run only on propane. The existing carburetor, fuel pump, and fuel tank are replaced with LP parts as described in Chapter 5. With a dual-fuel conversion, the gasoline fuel system components are not removed or modified. The LP mixer is mounted over the gasoline carburetor, and the rest of the parts are installed as an auxiliary fuel system so that the vehicle can run on either propane or gasoline.

Although either approach allows a vehicle to operate on propane, there are advantages and disadvantages that make straight conversion the best choice for some applications and dual-fuel conversion the best choice for others. Generally, straight conversions are most advantageous for those who have ready access to propane fuel supplies. Dual-fuel conversions are best for those who have to go out of their way to buy propane or who want the assurance of gasoline as a backup fuel in case they run out of propane in an area where they don't know where to buy it.

STRAIGHT CONVERSIONS

Straight conversions seem to be the most popular choice among fleet operators who make the switch to propane (Fig. 6-1). Fleets with a dozen or more vehicles will often set up a propane refueling station at the fleet's central garage. Because the vehicles

Table 6-1. Comparison of Straight and Dual-Fuel Conversions.

	Straight LP	Dual-Fuel
Conversion cost	About the same for both $800 to $1500	
Fuel economy	The same to 10 percent fewer mpg	Usually 10 percent to 20 percent drop in mpg on propane due to tuning compromises
Emissions	Low	Low when using propane
Performance	Good because the engine can be tuned to propane	May be 5 percent to 10 percent less when using propane because of tuning compromises
Weight penalty	None or some depending on fuel tank size	Usually some because vehicle must carry two fuel systems
Engine modifications	Might be necessary on older engines to install hard valves. Also involves ignition cooling & emissions modifications.	Usually no internal modifications made. Ignition modification minimal.
Best applications	Fleets or those with ready access to refueling facilities	Private vehicles where refueling may occasionally present problems.

usually return to the same garage every night, refueling presents no problems. Even for vehicles that travel long distances, refueling can be no problem if refueling stations are established at convenient intervals.

For example, suppose that a fleet operates a number of light delivery trucks between two cities. If the cities are further apart than the round-trip driving range of the fuel tank, the fleet owner could set up a refueling station in the other city or locate a local propane dealer in that city who would refuel the vehicles for the fleet.

Straight conversions can also be a workable alternative in many farming applications. A tractor or pickup truck that is used to haul feed and supplies over limited distances could be converted to straight propane. The fuel supply is usually no further away than the on-farm LP storage tank or the local propane dealer. The main advantages of choosing a straight conversion over dual fuel are as follows.

Better Performance. The engine can be tuned for optimum performance on propane without making any compromises for gasoline. Better fuel mileage, faster acceleration, smoother running and faster starting are what result from changing the ignition timing, emissions hardware, and fuel system to match propane's burning characteristics. When the LP mixer is stacked over the gasoline carburetor in a dual-fuel conversion, the added air restriction usually cuts horsepower by 5 percent to 8 percent.

Lower Driving Costs. Because all use of gasoline is eliminated, the cost advantage of propane is maximized.

Faster Payback. Although the costs of a straight conversion are about the same as a dual-fuel conversion, the straight conversion will pay for itself faster because the fuel savings are greater. The payback period for the dual-fuel setup depends on how often gasoline is used. If a vehicle is driven half the time on gasoline and half the time on propane, its payback period will be at least twice as long as the vehicle that runs on propane all the time.

Lower Maintenance. Because there is no gasoline to contaminate the oil or spark plugs, service intervals can be stretched further than they can in dual-fuel conversions. In addition, there's no gasoline carburetor, fuel filter or fuel pump to service or repair.

Reduced Weight Penalty. Vehicles with dual-fuel conversions carry the weight of two complete fuel systems. If both fuel tanks are full, the added weight can add up to several hundred pounds. Excess weight drags down fuel mileage.

No Fuel Pilferage. If there's no gas tank, it's pretty hard for anyone to steal gas. Although it is not impossible to drain an LP tank, a would-be thief would have to know what he's doing as well as have a propane-powered vehicle of his own in which to use the stolen fuel.

Lower Emissions. Because no gasoline is burned and because the engine can be tuned for optimum performance on propane, exhaust emissions will be substantially lower.

In spite of all the advantages, straight propane conversions aren't for everyone. A large percentage of the vehicles converted to propane each year are dual-fuel conversions because propane isn't as available as gasoline. You can't pull into your neighborhood gas station and fill up your tank with propane like you can gasoline. Because of this, many people want and need the assurance of having gasoline as a backup in case they find themselves running low on propane when there's no propane dealer nearby.

DUAL-FUEL CONVERSIONS

Dual-fuel conversions allow the flexibility of running on either propane or gasoline (Figs. 6-2 through 6-4). Under normal cir-

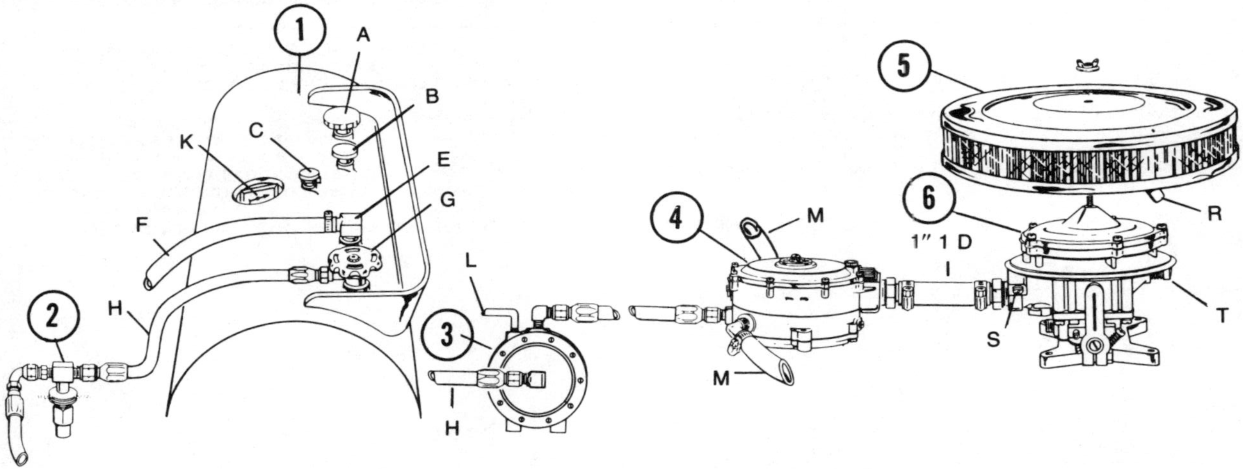

1. **LP-GAS MOTOR FUEL TANK** (A) Filler valve (B) Vapor return valve (C) 10% Outage valve (E) Relief valve (F) Vent line to outside of vehicle (G) LP-gas valve (H) LP-gas high pressure hose line (K) Fuel gauge.
2. **SS802 HYDROSTATIC RELIEF VALVE** Vented to outside of vehicle.
3. **VFF30 VACUUM FUELOCK AND FILTER** Prevents flow of fuel when engine stops with ignition switch on. Mount in a fire protected location (H) LP-gas inlet line. (L) Connects to air valve vacuum port in mixer.
4. **IMPCO MODEL EBP CONVERTER** Two stage regulator and vaporizer (M) Water inlet or outlet from engine. Must be brass fittings.
5. **IMPCO AF1-7 AIR CLEANER** Low profile design, dri-type PCV Connection.
6. **IMPCO MODEL *CA225 CARBURETOR** *Accepted by California Air Resources Board—Resolution No. 70-9A (T) Idle adjustment screw (S) Balance line connection

Fig. 6-1. Basic components of a straight propane conversion.

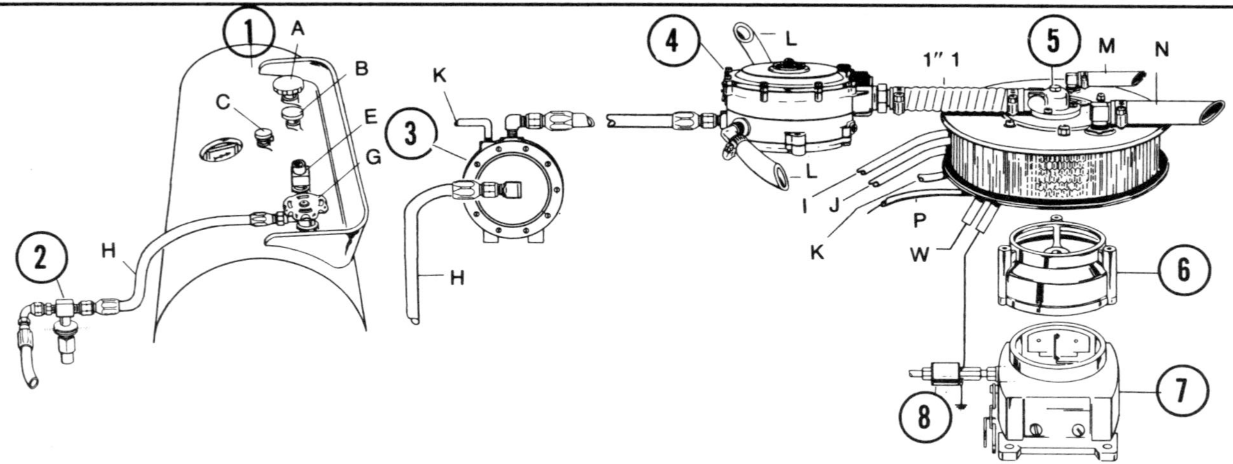

1. **LP-GAS MOTOR FUEL TANK** (A) Filler valve (B) Vapor return valve (C) 10% Outage valve (E) Relief valve (G) LP-gas valve (H) LP-gas high pressure hose line
2. **SS802 HYDROSTATIC RELIEF VALVE** Vented to outside of vehicle.
3. **VFF30 VACUUM FUELOCK AND FILTER** Prevents flow of fuel when engine stops with ignition switch on. (H) LP-gas inlet line. Vacuum line (K) connects to port provided in mixer.
4. **MODEL EOP LP-GAS CONVERTER** Two stage regulator and vaporizer (L) Water inlet or outlet from engine cooling system (Use Brass Fittings).
5. **CA300A-20 MIXER** (M) PVC connection. (N) Evaporative control vent (I, J) EGR Hoses. (K) VFF30 Vacuum Hose. (W) to ignition switch 12 volt source.
6. **ADAPTOR TO GASOLINE CARBURETOR** Supplied by IMPCO as specified. Straight or offset adaptor available.
7. **EXISTING GASOLINE CARBURETOR**
8. **GASOLINE SOLENOID VALVE** Operates from Micro Switch on Mixer.

Fig. 6-2. Basic components of a dual-fuel conversion.

Fig. 6-3. One of the drawbacks of dual-fuel conversions is the added weight and bulk of two complete fuel systems. This dual-fuel Jeep has lost much of its cargo space in the rear because of the vapor box around the propane fuel tank.

cumstances, most people use propane as the primary fuel and gasoline only as a backup for emergencies or to extend their driving range.

The biggest drawback of a dual-fuel system is the loss of performance that results from trying to tune an engine to run on two different fuels. If the engine is tuned for gasoline, performance will be sluggish when it's run on propane. If tuned for propane, it can knock and ping and be hard to start on gasoline. If a compromise is made and the engine is tuned at some point half way in between, performance will be less than its best on both fuels.

The solution most often used is to tune the engine for the fuel that will be used the most, and to accept the fact that performance will be less than desired when you are using the backup fuel. Because propane is the less expensive fuel and the reason for making the conversion in the first place, it makes sense to tune the engine as best as one can for propane. Unfortunately, compromises must still be made to avoid detonation, starting problems, and emissions problems with gasoline. This isn't to say the engine won't run satisfactorily on propane in dual-fuel conversions, but compared to the straight conversion there will be a noticeable difference.

Another reason for a loss in performance in dual-fuel conversions is the added air restriction caused by stacking the LP mixer over the gasoline carburetor. The second carburetor will restrict the amount of fuel drawn into the engine. At idle and low speeds the difference is negligible. At higher rpm's, the power loss can be as great as 5 percent to 8 percent.

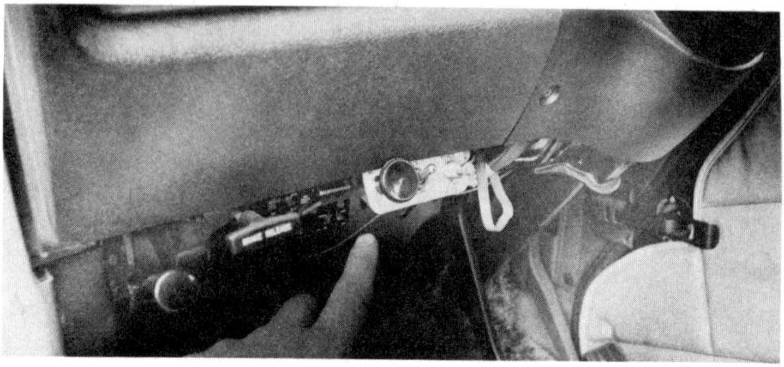

Fig. 6-4. A dual-fuel control switch under the dash controls fuel operation. Mastering the proper switching technique is important to prevent flooding the engine with fuel.

In a vehicle with marginal power to begin with (as is the case in many late model cars, pickups, and motorhomes), the loss in performance can be enough to hinder passing, hill climbing, and towing. If you're having trouble keeping up with traffic now, you may want to think twice about a dual-fuel conversion.

DUAL-FUEL OPERATION

Switching from propane to gasoline and back again is usually accomplished by manipulating a dash-mounted control handle and switch. The exact mechanisms vary from manufacturer to manufacturer, but most use a simple pull handle and toggle switch. The handle goes in for propane and out for gasoline. The engine can be switched from one fuel to the other at any time, with the engine running or with the engine off. It's a fairly simple procedure that takes little practice to master.

The pull handle is connected to the air valve on the LP mixer. The handle or a toggle switch is also wired to the propane fuelock and an electrically operated fuel shut-off solenoid in the gasoline fuel line. Pulling the handle out opens up the LP mixer air valve so that air bypasses it and flows directly into the carburetor. The fuelock shuts off the flow of propane and the solenoid on the gas line opens. This allows gas to flow into the careburetor.

Pushing the handle back in reverses the process. The LP air valve closes and causes the air to flow through the mixer before it enters the carburetor. The gasoline solenoid shuts off the flow of fuel to the carburetor and the fuelock opens so that propane can flow to the mixer. In other words, when the handle is shoved in the propane fuel system is opened up and the gasoline system shut down. When the handle is out, the propane system is shut down and the gasoline system opened up. In actual use, the procedure for changing from one fuel to the other is a little trickier than just pushing or pulling on the control handle and flipping a switch.

Switching from propane to gasoline is straightforward. Just pull the handle out. The propane is cut off immediately and within a few seconds gasoline is flowing into the carburetor. There is a brief lag while the engine goes from one fuel to the other, but it rarely causes any problems.

Going the other way—that is from gasoline to propane—requires an extra step and a little precaution. When switching from gasoline to propane, you must wait a few seconds between shutting off the gasoline and starting the propane to allow the engine to draw the remaining gasoline out of the carburetor. You do this by pushing the control handle only half way in and waiting until the engine starts to lose power. Shoving the handle half way in only shuts off the gas; it does not open up the propane system. It takes a couple of seconds for the engine to use up the small amount of gasoline that remains in the carburetor bowl. If you don't allow a few seconds for this to take place and just shove the handle all the way in, the engine will get too much fuel. Because it can't handle both gasoline and propane at the same time, it will probably stall.

Stalling is only a problem if you don't give the engine enough time to make the transition. You can minimize the risk of stalling by making sure you don't try switching from one fuel to the other when idling or driving at slow speeds.

To restart a stalled engine that died because of too much fuel, simply pull the control level half way out, hold the accelerator pedal to the floor, and crank the engine for a few seconds to blow the excess fuel out of the cylinders. In the half-way position, the engine will be getting no fuel (some people use this trick as an anti-theft device). After the excess fuel is out of the engine, pull or push the handle to start the engine on gasoline or propane.

Although the switch from one fuel to the other is not made all that often—and most drivers have little difficulty learning the technique—you should be prepared for the momentary lag that takes place during transition. Don't try to switch fuels while you are creeping along in heavy traffic or when you are climbing a long hill because you're apt to find yourself in a bind. Keep in mind that a stalled engine can be hazardous because you would lose your power-assisted steering and brakes. Use common sense and learn the proper technique for switching from one fuel to the other.

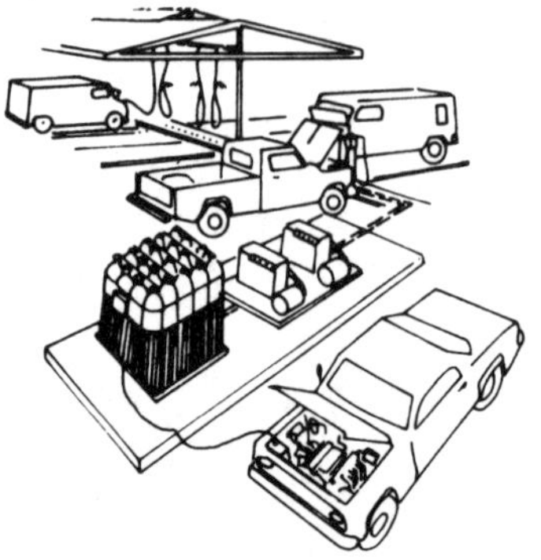

Chapter 7

Straight Propane Conversion

In this chapter, I will cover the basics of a straight propane conversion. The step-by-step procedures should serve only as a guide. Some mechanics will prefer to follow the conversion steps in their own sequence. One mechanic might choose to install the underhood components first and then the fuel tank. Another mechanic might do just the opposite. Many conversion centers and fleets use a two-man team to cut the conversion time in half. One man does the fuel tank installation while the other works under the hood on the carburetor and related hardware.

Regardless of the exact procedure used, the basic idea is to remove all the gasoline fuel system components and replace them with LP components. One part of the conversion that should follow the sequence outlined in this chapter is that of making and testing the final fuel line connections. It's important to make sure all pressure connections are gas tight, and that the various safety valves and fuel shutoffs are doing their job. When dealing with a pressurized fuel system, safety can't be overemphasized.

PRELIMINARIES

Before you begin the conversion process, there are a number of preliminary checks that should be made on the vehicle to determine the overall health of the engine (Fig. 7-1). It doesn't make much sense to spend $1000 to convert an engine that will soon require extensive and costly repairs. Before you touch anything, do

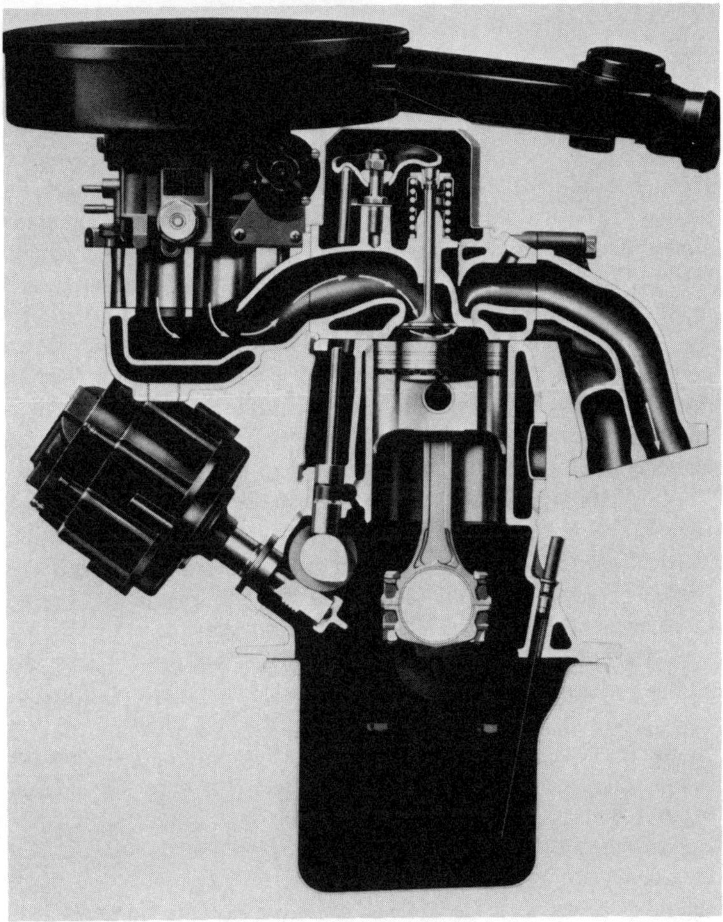

Fig. 7-1. Before the engine is converted to run on propane, it should be thoroughly checked to make sure it has good compression and is in relatively good condition.

a compression test to see that the rings and valves are in good condition.

Remove all the spark plugs from the engine and wedge the gas pedal linkage to hold the carburetor throttle wide open. Then install your compression gauge into one of the spark plug holes and crank the engine for about five seconds. Write down the compression reading, and then move on to the next cylinder and repeat the test. When you've finished reading the compression in each cylinder, the numbers should not vary by more than 20 percent from one cylinder to another.

Most engines in good condition will show compression readings between 110 and 150 psi. If the readings are substantially lower, say down around 70 to 90, the rings or the valves are worn and should be repaired. To determine whether the rings or valves are to blame, pour about an ounce of engine oil into a cylinder through the spark plug hole. Turn the engine over a couple of times so that the oil will temporarily seal the rings. Repeat the compression test. If the compression increases over the earlier readings, the rings are at fault. If there's little or no change, the valves are leaking and should be reground.

A very low compression reading in two adjacent cylinders is a good indication of a blown head gasket. A zero compression reading in any one cylinder usually indicates a bent, stretched, or damaged valve. Whatever the cause, any compression problems should be corrected before converting the engine to propane.

If the engine passes the compression test, has good oil pressure, is not making any unusual noises that could indicate bearing, rod or lifter trouble, and is not using oil (an indication of worn valve guides or seals), you can safely assume the engine is in good mechanical condition and is suitable for conversion.

Next on the check list is the ignition system. To fire the air/fuel mixture, the spark plugs, ignition cables, distributor cap, rotor, ignition coil, contact points or electronics must be in top condition. A weak spark will cause misfiring, loss of performance, reduced fuel economy and hard starting. Make sure the ignition system is up to specifications before converting to propane.

If the spark plugs have more than 20,000 miles on them, or if they are coated with heavy deposits or show signs of wear, they should be replaced. If the ignition cables have cracks, burned spots, loose fitting terminals or have high internal resistance (no more than 30,000 ohms for cables up to 25 inches in length or 50,000 ohms for longer cables), you'll need a new set of cables. The distributor cap should be free from cracks, carbon tracks, or corroded terminals. The rotor must also be free from carbon deposits and oxidation. Pitted or corroded contact points should be replaced. Any problems in an electronic ignition system should be solved before making the conversion to propane.

One way to quick check the health of the ignition system is to pull the coil cable from the center of the distributor cap and hold it about ¼ of an inch from the engine (any grounded metal surface will do). Turn the ignition switch to the "on" position and have a helper crank the engine over a couple of times (you should be using insulated pliers or gloves to avoid being shocked). If the ignition system is in good working condition, the spark will jump the gap with a sharp cracking noise, and you should see a bright white or blue spark. If there's no spark or a weak yellow spark, there's a problem in the ignition system that must be corrected. A weak coil or worn or misaligned contact points are the most likely culprits.

After checking out the ignition system and making any necessary repairs to bring it up to par, inspect the starting and cooling systems. Check the battery, battery cables, and starter to see that all are working as they should. Then check the radiator, hoses, belts, water pump and fan to see that there are no surprises waiting to cause trouble.

Because you'll be installing a colder thermostat as part of the conversion, now might be a good time to drain and flush the cooling system. When you refill the radiator after installing the new thermostat, use fresh antifreeze if the coolant is more than two years old. Always use a 50/50 mixture of antifreeze and water. Straight water offers absolutely no corrosion protection, no freezing protection, and limited boilover protection.

Many problems often attributed to propane are usually due to conditions that existed before any modifications were ever made. If an engine has a weak ignition system, borderline cooling capacity, or internal problems, those problems will make themselves known before long. Propane will probably be blamed. Unfortunately, conversions are often made rather hastily with little or no attention given to checking out the basic health of the engine and correcting any problems that are uncovered. So save yourself needless grief and anxiety by taking the time to thoroughly inspect the engine, the ignition, the starting system and the cooling system as part of your conversion process. The results will be well worth the effort.

LP-EQUIPMENT SELECTION

Once you've completed a basic inspection, you're ready to select the LP equipment for your conversion. Don't buy anything that doesn't meet ASME, DOT, or UL standards. All of the major propane equipment manufacturers build their components to meet accepted industry standards. Beware of the "do-it-yourself" kits or plans that call for recycled propane bottles or other makeshift equipment. Although such equipment might appear to work satisfactorily, it can be extremely dangerous because it lacks the built-in safety features found in approved equipment. You'd be foolish to risk your life and the lives of others just to save a few bucks on the conversion cost.

The two pieces of equipment that should be selected with some degree of care are the fuel tank and the carburetor. When deciding on what size fuel tank to buy, keep in mind the intended use of your vehicle (Table 7-1). If the vehicle is to be used for short trips only, you certainly don't need to buy a 200-gallon tank. A 30- or 40-gallon tank should work fine. On the other hand, if you plan on doing a lot of driving between fill-ups or if you don't have easy access to refueling, a larger tank would be the wiser choice.

When you are estimating what size tank you'll need, keep in mind that your fuel mileage will probably drop about 10 percent when you convert to propane. If your vehicle presently gets around 15 mpg, you'll get 13 mpg to 14 mpg on propane. If you'd like to have a driving range of 400 miles on a tankful of fuel, you'd need a tank that holds 30 gallons of fuel. Because LP tanks are never filled beyond 80 percent of capacity, you'll actually need a tank that's 20 percent larger than the number of gallons you want to carry. In this example, you'd need a 36-gallon tank if you wanted a tank capacity of 30 gallons.

Because the fuel tank accounts for a large portion of the conversion expense, it's best to give the matter some thought before you buy anything. Don't buy a bigger tank than you really need, but make sure the tank is big enough to give your vehicle a practical range. Nothing is more frustrating than to spend $1000 on a conversion only to discover several months later that you should have bought a bigger tank.

The other piece of equipment that must be "sized" to the application is the carburetor. A carburetor that's too small for a given engine will limit horsepower. At higher rpm, an undersized carburetor will restrict the air flow and prevent the engine from developing its full power. A carburetor that's too large for a given engine can also cause problems. If there's too much throat area, the air will not move with sufficient velocity as it passes through the carburetor. Low rpm torque and throttle response suffers, and starting can be a problem.

Generally, the rated air flow capacity of the carburetor should be matched to the air flow requirements of the engine. In other words, big engines usually require big carburetors whereas little engines need little carburetors.

Table 7-1. Estimated Driving Ranges for Various-Sized LP Fuel Tanks.

LP Fuel Tank Capacity (in gals. of water)**	Vehicles mpg on gasoline/propane* 5/4.5 mpg.	10/9	15/13.5	20/18
20 gallons	72 miles	144	216	288
25	90	180	270	360
30	108	216	324	432
35	126	252	378	504
40	144	288	432	576
50	180	360	540	720
60	216	432	648	864
80	288	576	864	1152
100	360	720	1080	1440
150	540	1080	1620	2160
200	720	1440	2160	2880

* First number represents vehicles mpg on gasoline. Second number shows a 10 percent drop in mileage when running on propane.

**Actual fuel capacity of an LP tank is 80 percent of the water capacity as room for fuel expansion must be allowed. This factor has been taken into account when computing the driving ranges.

Engine speed and use are also important considerations when it comes to picking the right sized carburetor for a given application. If an engine is always run at low rpm and part throttle, a smaller carburetor can be used than the one that would be required for an identical engine that is run at high rpm and full-throttle.

Table 7-2 shows the engine air flow requirements for various sized engines at different rpm. To determine the best carburetor for your vehicle, find the cubic inch displacement (CID) of your engine in the left hand column. Then read across to the rpm column that best describes the *maximum* operating rpm of your engine. The number at that point represents the cubic feet of air per minute (cfm) your engine requires to develop full power. That's the size carburetor you should install on your engine.

For example, suppose that you have a 350 cubic CID Chevy pickup that would probably never exceed 4000 rpm in everyday use. The maximum rated cfm requirement for that engine would be 344. A carburetor rated at 350 cfm would satisfy the air requirements of the engine for normal use.

For a heavy-duty 400-CID truck engine that is governed to run at speeds below 3000 rpm, the maximum cfm requirement of the engine would be 295 cfm. Here a carburetor rated at 300 cfm would be satisfactory.

Table 7-2. Engine Airflow.

Over-the-road vehicle applications

CID	RPMs 2500	3000	3500	4000	4500	5000	5500	6000
100	62 cfm	74	86	98	111	123	135	148
150	93	111	129	148	166	185	203	221
200	123	148	172	197	221	246	271	295
250	154	185	215	246	277	307	338	369
300	185	221	258	295	332	369	406	443
350	215	258	301	344	387	430	474	517
400	246	295	344	394	443	492	541	590
450	277	332	387	443	498	553	609	664
500	307	369	430	492	553	615	676	738
550	338	406	474	541	609	676	744	812

Industrial (stationary and portable) applications

CID	RPMs 500	750	1000	1250	1500	2000	2500
500	62 cfm	92	123	154	185	246	307
550	68	102	135	169	203	271	338
600	74	111	148	185	221	295	369
650	80	120	160	200	240	320	400
800	98	148	197	246	295	394	492
1000	123	185	246	307	369	492	615

The CFM (Cubic Feet per Minute) airflow requirements of a given engine can be determined by finding the CID (Cubic Inch Displacement) in the left hand column and then reading across under the RPM (Revolutions Per Minute) column that represents the maximum operating speed of the engine.

When choosing an actual carburetor, it's best to pick one that has a little over-capacity than one that doesn't quite satisfy the maximum air flow requirements. For example, a 351 CID Ford engine at 4000 rpm needs about 344 cfm of air. If you're looking at Impco carburetors, you'll have to choose between a Model 225, which is rated at 329 cfm, and a Model 425 which is rated at 460 cfm. If there's a good chance that the engine will be operated at or near 4000 rpm periodically, it's best to go with the bigger Model 425 carburetor. On the other hand, if the engine is used in such a way that it rarely approaches 4000 rpm, the smaller carburetor would be satisfactory.

STEP-BY-STEP CONVERSION PROCEDURES

With the preliminary inspection completed and the properly-sized LP equipment selected, the vehicle is ready to be converted. The following steps describe the basics of what should be done. I want to stress that this is not a job for the novice mechanic. Unless you have good mechanical skills and knowledge, and have had previous experience or training in this area, the conversion is best left to someone who knows what he's doing.

Remove the air cleaner assembly complete with its heated air ducting and vacuum plumbing (Fig. 7-2). Because the stock air cleaner is usually replaced with a low restriction air cleaner, the stock unit can be discarded. The vacuum line to the warm air trap door should be plugged because this piece of emissions control hardware is no longer needed.

Disconnect the throttle linkage, transmission kick-down linkage (automatics only), fuel lines and vacuum hoses from the carburetor. Label each if necessary to aid reassembly.

Unbolt the carburetor from the intake manifold and remove the carburetor (Fig. 7-3).

If the engine is equipped with an exhaust gas recirculation (EGR) valve, remove or plug the vacuum control line to render it inoperative. On some engines, the EGR valve can be removed entirely.

If the engine is equipped with an air pump, it can be removed along with the lines to the exhaust manifold(s). Plug the openings in the exhaust manifold(s) with pipe plugs. If the catalytic converter is to be retained, the air pump should be left intact.

Do *not* remove the PCV valve; it is essential to proper crankcase ventilation. A new PCV breather connection will be provided on the replacement air cleaner.

Fig. 7-2. The stock air cleaner and warm-air snorkel assembly is removed and discarded.

Fig. 7-3. The gasoline carburetor and fuel lines are removed.

Remove the fuel line that runs from the fuel pump to the carburetor, and then remove the fuel pump. Use a block-off plate to seal the fuel pump mounting hole on the engine block. This is necessary to prevent oil from leaking out the pump opening (Fig. 7-4).

Remove the throttle plate from the bottom of the carburetor (Fig. 7-5) and mount the plate on the bottom of the propane carburetor using the required adapter or gaskets. Install the linkage adapters on the carburetor assembly as required.

Mount the propane carburetor on the engine using a new gasket. Use gasket sealer on the gasket to prevent vacuum leaks from forming. Tighten the carburetor down carefully and evenly to avoid distorting the base.

Reconnect the throttle linkage, transmission linkage (automatics only), and vacuum lines to the carburetor.

Determine the best location for the converter and fuelock (Fig. 7-6). Both should be located away from the exhaust manifold(s). Both should also be located in reasonably accessible positions. For optimum coolant flow through the converter, it should be located at a point lower than the top of the radiator.

Install the mounting brackets for the converter and fuelock. Install the fuel line fittings, vacuum fittings, and coolant fitting on the converter and fuelock. Next bolt each to its mounting bracket (Fig. 7-7). Only brass coolant fittings should be used on the converter to prevent electrolysis corrosion of the aluminum housing. All fittings should also have sealer applied to the threads to prevent fuel leaks, vacuum leaks, or coolant leaks.

Make the coolant connections for the converter by tapping into the heater hoses. The connection must be made so that warm water will flow through the converter whether or not the heater is on (Fig. 7-8). Stainless steel worm-type clamps are recommended for all coolant-hose connections.

Fabricate a fuel line and fittings for the fuelock to the converter using braded pressure hose and screw-on fittings (Fig. 7-9). Apply sealer to the fitting threads and install the fuel line.

Make the necessary vacuum or electrical connections for the fuelock. Follow the equipment manufacturer's instructions.

Remove thermostat housing and install a 160 or 170 degree "summer" thermostat. Apply sealer to a new gasket before replacing the thermostat housing.

Fig. 7-4. The gasoline fuel pump is removed and a plate is installed over the mounting hole on the engine block.

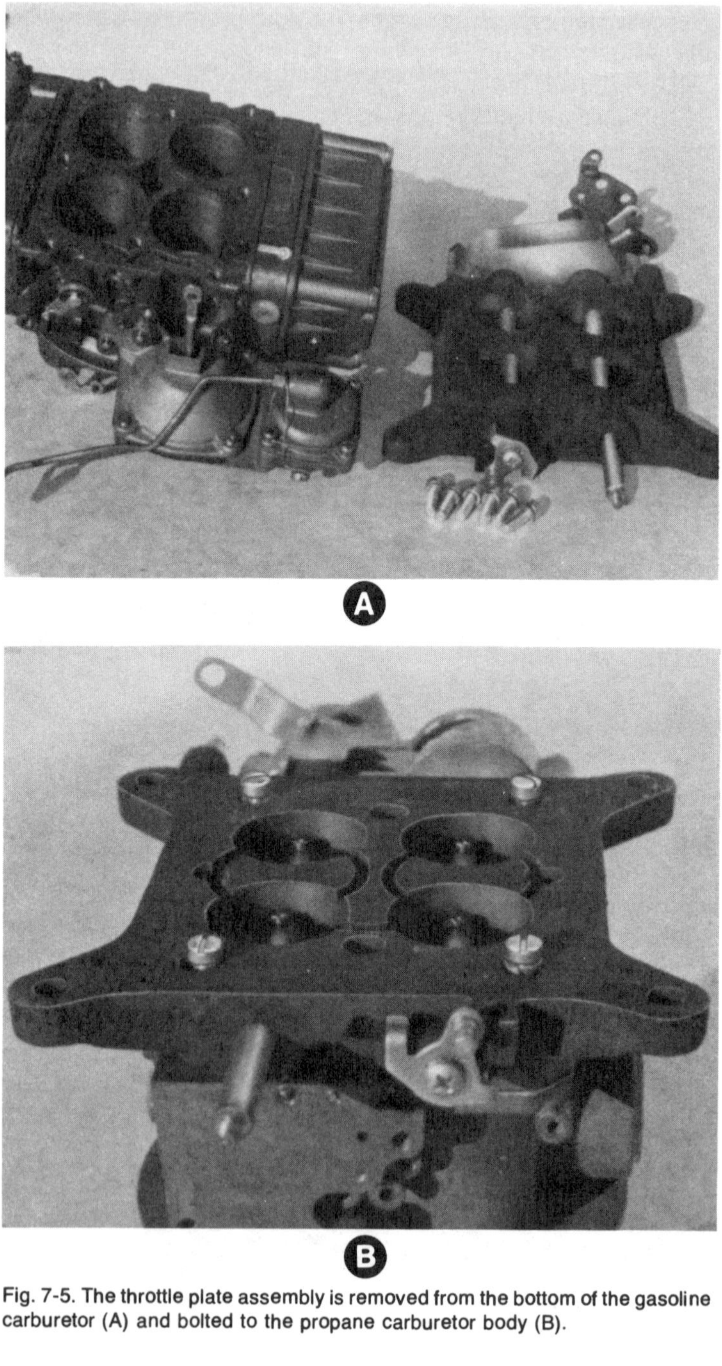

Fig. 7-5. The throttle plate assembly is removed from the bottom of the gasoline carburetor (A) and bolted to the propane carburetor body (B).

Fig. 7-6. The converter is positioned in the engine compartment for easy access and away from the exhaust manifold.

If colder spark plugs are to be used, install them now (Fig. 7-10). Plugs should be gapped no wider than .028 inches on a point ignition system, or .035 inches on a high energy electronic ignition system.

The distributor can be recalibrated by removing it from the engine and recalibrated on a distributor machine, or an advance kit consisting of heavier weights, lighter springs, and a limited bushing can be installed in the distributor while it's still in the engine (Fig. 7-11). To install the kit, remove the distributor cap and rotor on non-High Energy Ignition GM distributors to gain access to the advance mechanism. On others, the base plate under the contact points or pickup module must be removed to get at the advance mechanism.

Connect the mechanical or electrical primer on the converter, and install the dash control switch or knob (Fig. 7-12).

On turbocharged applications or situations where a high air intake restriction will exist, a balance line must be installed from the converter and fuelock vent openings to the air horn on the carburetor.

The gasoline fuel tank can now be removed. Disconnect the fuel line and drain the tank. Then unbolt the hangers and lower the

Fig. 7-7. A converter and vacuum fuelock installed on the fenaerwell of the engine compartment.

tank from the vehicle. If the gas tank is to be left in place, make sure the fuel outlet line is plugged. A quart or two of 10-weight oil poured in the tank will keep it from rusting while it is not being used.

Determine the best location for the LP fuel tank and make a trial fit to see how things line up (Fig. 7-13). Then install the tank

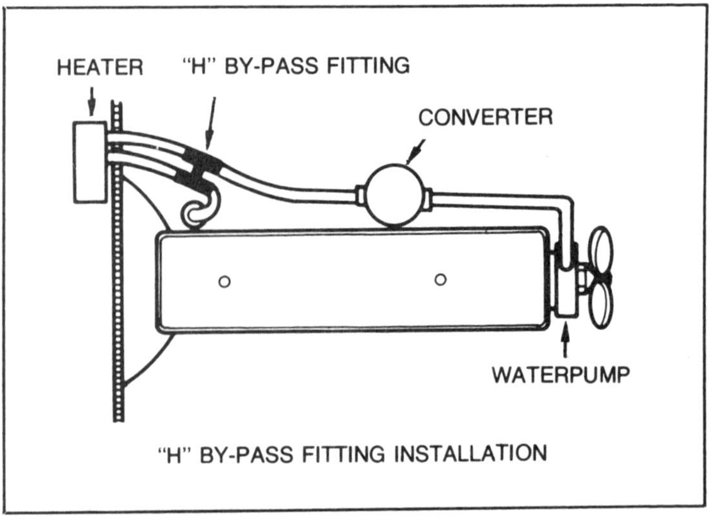

Fig. 7-8. Coolant connections for the converter.

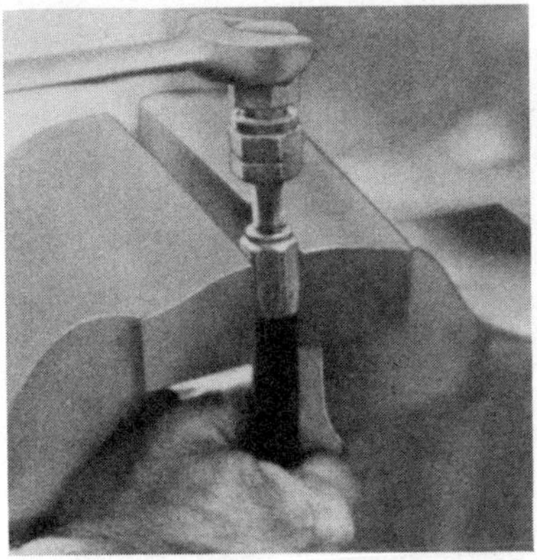

Fig. 7-9. Fabricating fuel-line connections using braded hose and screw-on aircraft-type fittings.

mounting brackets by drilling holes and bolting the brackets in place. The NFPA recommends using 7/16-inch bolts with ⅛-inch backing plates that provide at least 7 square inches of surface area. The backing plates prevent the bolts from pulling through the sheetmetal in case of an accident.

If the fuel tank is being installed in the trunk of a passenger car, a vapor seal must be formed to prevent LP fumes from entering the passenger compartment. Urethane foam can be used to seal the area between the back seat and trunk. All radio speaker openings or electronic equipment in the trunk must also be sealed. The trunk should also be vented to the outside by providing a 3-square-inch opening in the floor of the trunk at its lowest point. This allows fumes to escape and also allows air to escape when you slam the trunk shut. A small baffel plate will keep water and dirt out of the trunk.

Fig. 7-10. Spark plugs with a colder heat range are recommended for straight conversions. Spark gaps should be set narrower for propane.

Fig. 7-11. Distributor advance curve is recalibrated by changing the centrifugal advance weights and springs. Shown is a GM HEI distributor.

Fig. 7-12. A cable connection to the primer button on the converter allows the driver to prime the fuel system without raising the hood. Electrically operated automatic primers are also available.

Fig. 7-13. Dual LP tanks can be installed in the spare tire well of a station wagon. A vapor barrier must be built around the tank to prevent fumes from entering the passenger compartment.

Drill holes for the pop-off valve, and remove filler pipe and fuel lines as required.

Install the fuel tank on the mounting brackets and bolt it securely in place. Install the safety release valve, the spit valve, the manual shut-off valve, and the fuel gauge in the tank. Use sealer on all threaded connections to make a gas-tight seal.

Install the remote filler pipe, spit-valve line, and pop-off line as required (Figs. 7-14 through 7-18). Remote filler pipes, spit valves,

Fig. 7-14. The safety pressure release pop-off valve for the fuel tank must direct fumes outside the vehicle for internal-mounted tanks. On passenger cars, a hole is usually drilled in the area between the trunk and rear window.

Fig. 7-15. Here's what the pop-off valve looks like after it's installed. A duplicate pop-off valve is being held nearby for comparison.

and pop-offs are only required when a fuel tank is installed in the trunk of a passenger car or inside an enclosed compartment in a truck or van. If a topper is installed over the bed on a pickup truck, a remote filler pipe, a spit valve, and a pop-off valve must be installed.

Install the fuel line from the fuel tank to the engine compartment using the recommended braided hose and pressure fittings. A hydrostatic relief valve must be installed in the fuel line between the fuel tank and fuelock. The relief valve must be located away from any heat source (such as a hot exhaust pipe or muffler) and protected from grease and dirt.

Fig. 7-16. The LP tank remote filler valve can be located anywhere that is convenient. In this case, it's located behind the license plate next to the old gas tank cap. Note the spit valve at the right.

Fig. 7-17. The remote filler on this van is located next to the rear wheel. This type of installation is not recommended because the location offers no protection to the valve from road hazards.

Put a couple of gallons of propane into the fuel tank and open the manual shut-off valve for a few seconds to clear the fuel line of any dirt or debris. Do this outdoors and away from any open source of flame. Then shut the valve.

Connect the loose end of the fuel line to the fuelock (Fig. 7-19). Open the manual tank shut-off valve and check the lines for leaks

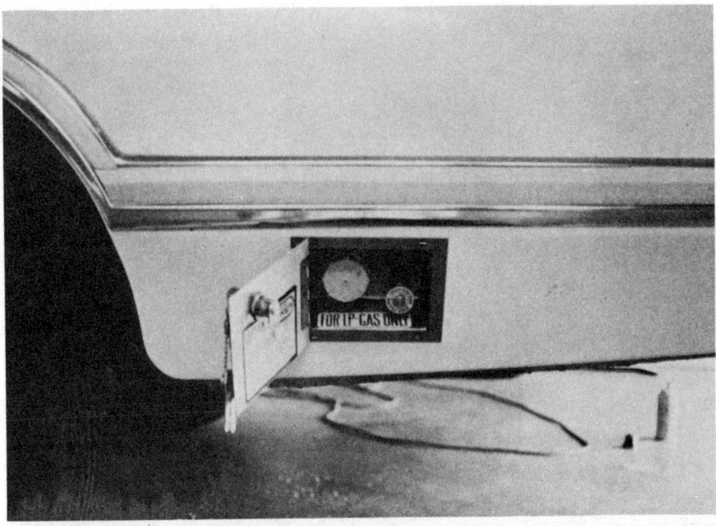

Fig. 7-18. The remote filler valve in this instance is well protected by the box and door surrounding it. This type of installation is common on converted pickups with bed-mounted tanks enclosed by toppers.

Fig. 7-19. After the main fuel line from the tank has been cleared by opening and closing the manual shut-off valve, the line can be connected to the fuelock.

with an electronic leak detector or soapy water. If soapy water is used, no bubbles should appear around any of the fuel line connections and the system should hold pressure. Leave the manual shut-off valve open.

Finish the fuel system installation by installing the vapor hose between the converter and carburetor (Fig. 7-20), installing the air cleaner (Fig. 7-21) and connecting the PCV breather hose.

Fig. 7-20. Connecting the vapor hose between the carburetor and converter.

Fig. 7-21. A conversion nears completion. After the air cleaner is installed, the engine can be started and adjusted.

Refill the cooling system with a 50/50 mixture of water and antifreeze.

Remove the heat riser valve from the exhaust manifold or wire it in the open position.

Fig. 7-22. After starting the engine, the ignition timing and carburetor must be adjusted. A chassis dynamometer and exhaust analyzer are highly recommended.

The engine should now be ready to start. Make a final check of all systems and connections, and then press the converter primer button to allow some propane to flow to the carburetor. Turn the ignition on and crank the engine until it starts.

As soon as the engine starts, adjust the idle speed and mixture temporarily so that the ignition timing can be set and checked (Fig. 7-22). Set the timing and check the advance with a timing light. Initial timing should be advanced several degrees over stock (with total advance coming in much earlier). Under no circumstances should maximum spark advance exceed 40 degrees on automotive engines, or 32 degrees on low-rpm truck engines.

After the engine reaches normal operating temperature, make the final adjustments to the idle speed and carburetor idle mixture. The vehicle should then be put on a chassis dynamometer so that the carburetor power adjustment can be set with the engine running under the load at road speed.

With the adjustments complete, fill the fuel tank with propane and recheck the system for leaks. Then road test the vehicle to check its performance.

The conversion process is now complete. The only other thing to do is to make sure that anyone who is likely to drive the vehicle knows that it is propane-powered. You don't want someone to try to fill the fuel tank with gasoline.

Chapter 8

Dual-Fuel Conversions

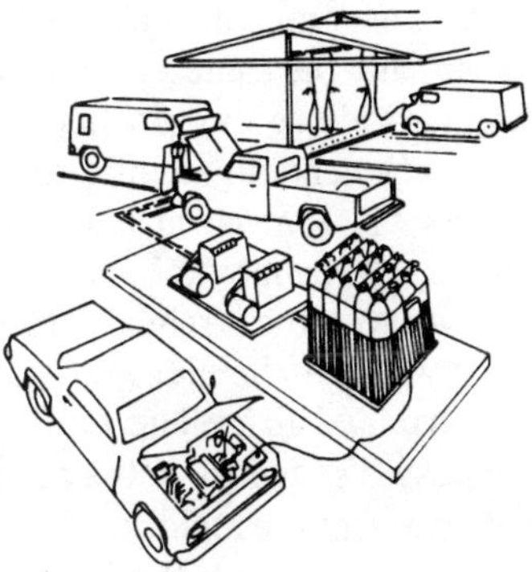

A dual-fuel conversion is different from a straight conversion in several important respects. First, the gasoline fuel system and its related emission controls are left intact. Second, the engine must be tuned so that it can run satisfactorily on either propane or gasoline. Ideally, it should be tuned to run the best on whichever fuel is to be used most often. Third, the propane mixer is installed on top of the gasoline carburetor. A fuel shut-off solenoid must also be installed in the gasoline fuel line as well as a master control switch and knob for switching from one fuel to the other.

As with a straight conversion, the mechanical condition of the engine should be determined before any money or time is invested in installing a propane dual-fuel system. If the engine and its ignition system, starting system, and cooling system all pass inspection, the next step is to select the proper sized fuel mixer for the application and a fuel tank. The propane tank should be selected according to the preferred driving range. Remember that the vehicle's total driving range will be the result of the LP tank plus the gas tank (assuming you start out with both tanks full).

One item that will require special consideration in a dual-fuel conversion is the added weight of the LP fuel tank. Because the gasoline tank is retained, the added weight of a full LP tank can add several hundred pounds to the chassis. For a passenger car (especially a small one) or a light-duty pickup truck, overload-shock absorbers, air-assist rear shocks, or overload springs might be necessary to maintain normal ride height and handling. Larger vehicles should have no trouble accommodating the added weight.

CONVERSION PROCEDURES

The following steps describe the basics of a dual fuel conversion. The tank, fuelock, fuel line, and converter installation is essentially the same as in a straight conversion. Only one modification is made to the gasoline fuel system (the installation of a fuel shut-off solenoid). Modifications to the ignition system and elimination of the exhaust manifold heat-riser valve depend on how frequently propane will be used as the primary fuel.

As with the straight-conversion procedure, it makes no difference whether the fuel tank or underhood fuel system components are installed first. The only part that should be followed in the suggested sequence is the purging of the main fuel line before it is connected to the fuelock. This minimizes the risk of passing rust or debris, that is inside the fuel tank or line, through to the fuelock. See Figs. 8-1 through 8-5.

Remove the stock air-cleaner assembly and its heated air ducting and vacuum plumbing. Because the stock unit will be replaced with the mixer air cleaner assembly, it can be discarded.

Trial fit the mixer air-cleaner assembly on top of the carburetor to check hood clearance. An offset air-cleaner adapter will be necessary in some vehicles for the hood to close properly (Fig. 8-6).

Trial fit the converter and fuelock in the engine compartment. Each component should be located away from the exhaust manifold(s), within easy reach, and where they will not obstruct or interfere with access to other frequently serviced items. Follow the equipment manufacturer's mounting instructions; some fuelocks and converters are designed to be installed in a certain position. For example, a fuelock that is supposed to be mounted vertically will not work properly if mounted in any other position.

Having determined the approximate locations for the converter and fuelock, the hose fittings can now be installed in these units. Coat the fitting threads with sealer to prevent leaks. Use only brass fittings for the converter hot-water connections because steel will cause electrolysis corrosion of the converter's aluminum case.

Install the converter and fuelock mounting brackets.

Mount the converter and fuelock on their brackets (Fig. 8-7).

Drain the cooling system and make the hot-water connections for the converter by tapping into the heater hoses. The connections

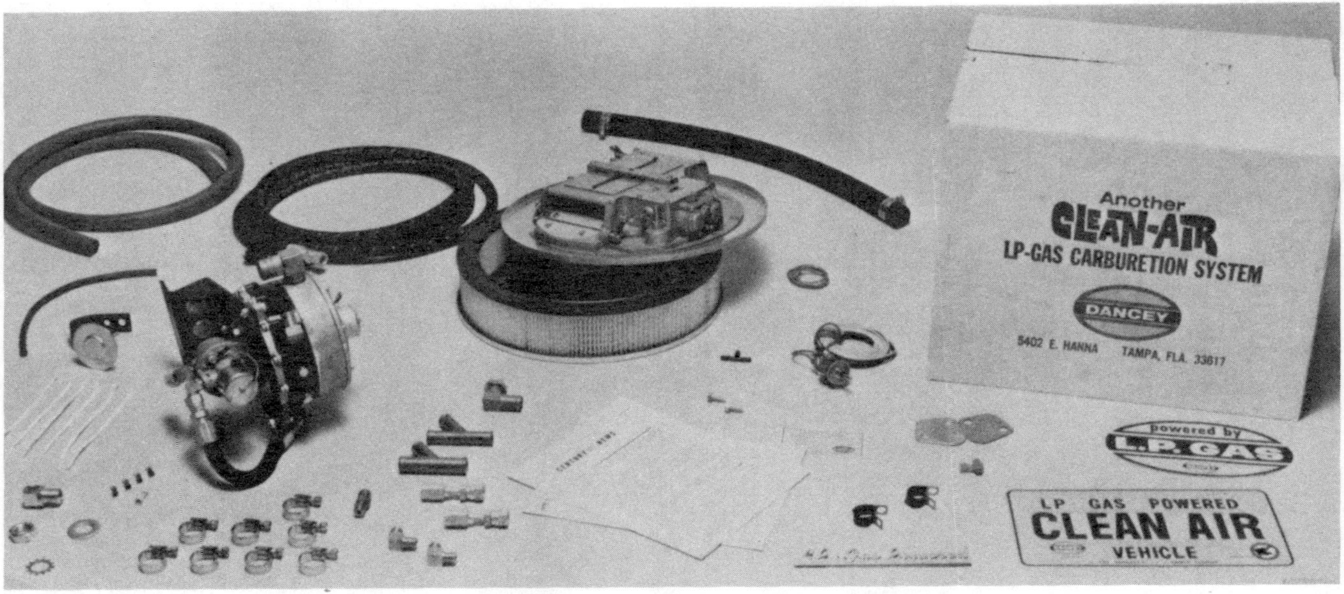

Fig. 8-1. A dual-fuel kit from The Dancey Company uses Acucarb carburetion and includes all the hardware necessary for a complete conversion. The fuel tank is ordered separately.

Fig. 8-2. A retrofit dual-fuel conversion kit is available through International Harvester dealers for IH trucks. The kit is packaged by Future Fuels of Bowling Green, KY. Acucarb carburetion is used.

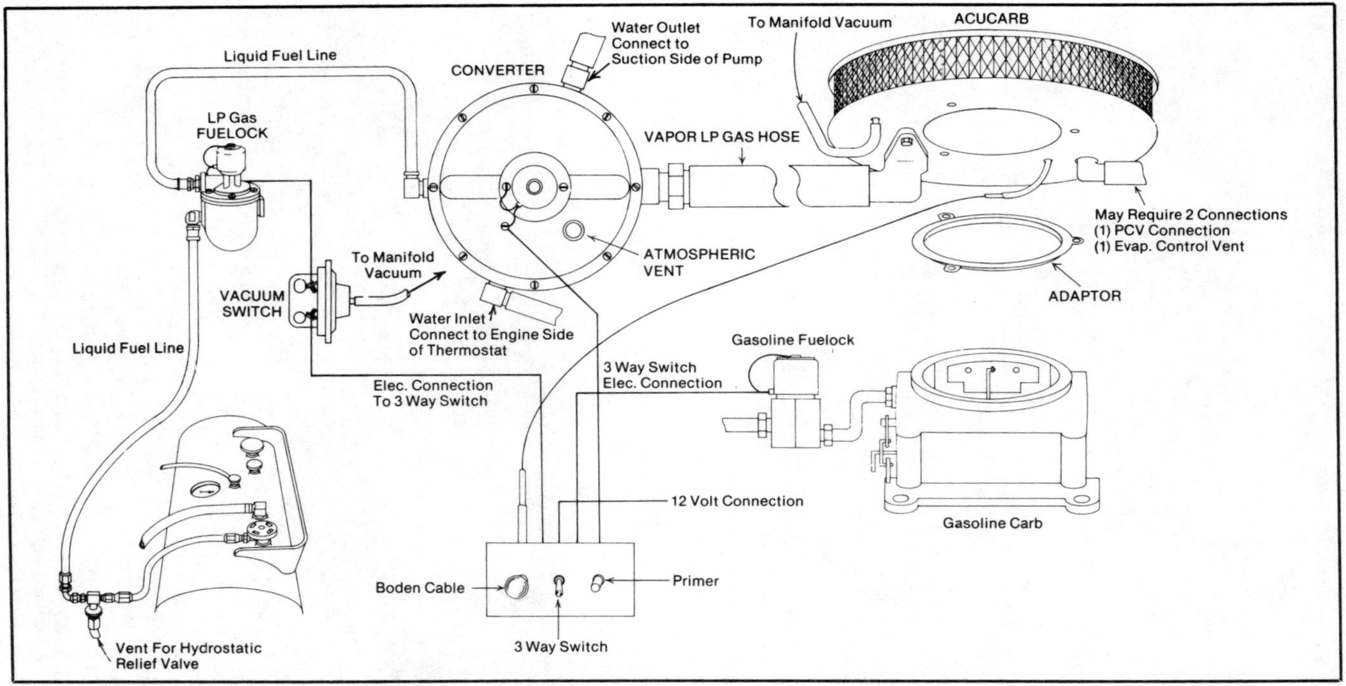

Fig. 8-3. A typical Acucarb installation.

Fig. 8-4. The Acucarb conversion installed on an IH truck.

must allow hot water to circulate through the converter whether or not the heater is on. The direction of flow is not important.

This step is optional. Remove the thermostat housing and install a 160- or 170-degree "summer" thermostat. This modification is recommended for any application where propane will be used as the primary fuel. For cold climates or for applications where propane will serve as the backup or secondary fuel, the stock thermostat can be used.

Fig. 8-5. An Algas dual-fuel conversion. The large black hose going to the air cleaner from the left is the propane vapor hose. The other black hose from the right is for the PCV breather.

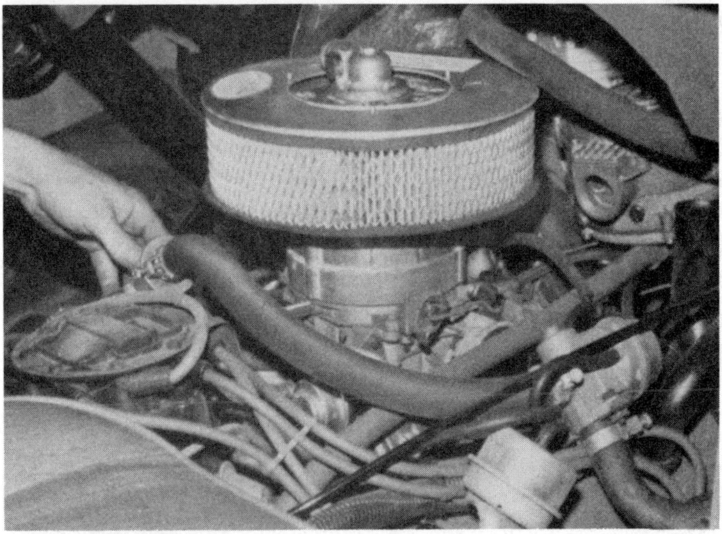

Fig. 8-6. Trial fitting the LP fuel mixer on top of the carburetor is necessary to check for hood clearance.

Refill the cooling system. If the old coolant appears rusty or muddy, dump it and refill the radiator with a fresh mixture of half antifreeze and half water.

Fabricate a fuel line for connecting the fuelock to the converter. Use only the recommended braded pressure tubing with

Fig. 8-7. Installing the converter and fuelock.

screw-on pressure fittings. Then install the line using sealer on the fitting threads.

Make the necessary vacuum or electrical connections for the fuelock. Follow the equipment manufacturer's instructions.

This step is optional. Install one-heat-range-colder spark plugs if propane will be used as the primary fuel. Otherwise, use the stock spark plugs. The only problem that can result from using the colder plugs is premature fouling when burning gasoline. If fouling is a problem on gasoline, switch back to the stock spark plug. If misfiring is a problem on propane, narrow the plug gap to .028 inches, on point ignition systems, or .035 inches on high-energy electronic ignition systems.

This step is optional. Remove the distributor and recalibrate the spark advance curve or install an advance curve kit with the distributor in the engine. This modification is not recommended for any application where gasoline will be used frequently. The over-advanced timing will cause detonation and pinging that could damage the engine. On most dual-fuel conversions, the basic ignition timing is simply advanced a couple of degrees and the advance curve is left stock. Some pinging might still result when running on gasoline, but it shouldn't be severe enough to cause any problems. It's just another one of the compromises necessary for a dual-fuel conversion.

Connect the mechanical or electrical primer on the converter.

Install the dash-mounted control panel, control cable, and electrical connections for the fuelock, converter, and gasoline solenoid.

Install the fuel shut-off solenoid in the gasoline fuel line between the carburetor and fuel pump (Fig. 8-8). Connect the electrical lead to the wire from the control switch. The solenoid should also be grounded to the engine block or intake manifold because grounding through the fuel line is sometimes inconsistent. Make sure the solenoid is installed in the proper position with the arrow indicating the direction of flow pointing towards the carburetor.

Make the necessary cable and electrical connections with the control mechanism on the fuel mixer. Follow the equipment manufacturer's instructions; one brand of mixer will differ from another.

Install the fuel mixer on the carburetor adapter.

Install the vapor hose from the converter to the fuel mixer. The engine compartment fuel system installation is now basically complete. Next the fuel tank and fuel line can be installed.

Determine the best location for the LP fuel tank and make a trial fit. Then install the tank mounting brackets by drilling holes

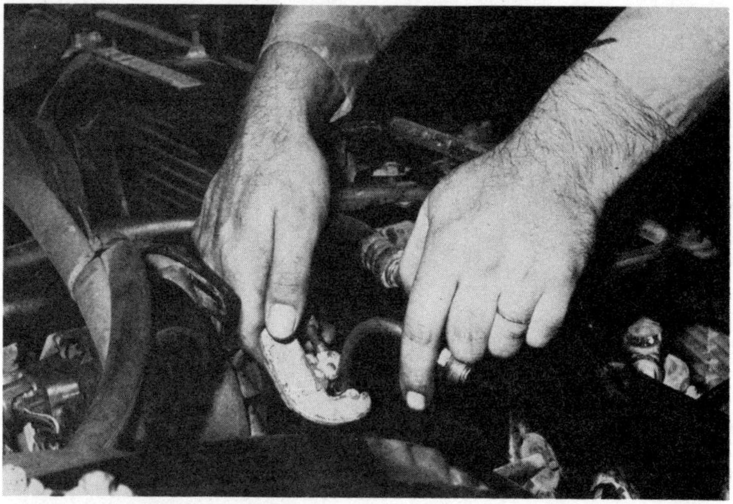

Fig. 8-8. To install the gasoline fuel shut-off solenoid, a small section must be cut from the fuel line.

and bolting the brackets to the vehicle's frame or chassis. When mounting to sheet metal, the NFPA recommends using 7/16-inch bolts with ⅛-inch backing plates that provide at least 7 square inches of surface area.

If the fuel tank is installed in the trunk of a passenger car, a vapor seal must be formed to prevent LP fumes from entering the passenger compartment. Urethane foam can be used to seal the area between the back seat and trunk. All radio speakers or other openings must also be sealed from the trunk side. The trunk should also be vented to the outside by providing a 3-square inch opening in the floor of the trunk at its lowest point. This allows fumes to escape and also allows air to escape when you slam the trunk shut. A small baffel plate will keep water and dirt out of the trunk.

Drill holes for the pop-off valve (Fig. 8-9), the remote filler pipe, and the fuel lines as required.

Install the fuel tank on the mounting brackets and bolt it securely in place. Install the safety release valve, the spit valve, the manual shut-off valve, and the fuel gauge in the tank. To avoid leaks, use sealer on all threaded connections.

Install the remote filler pipe, the spit-valve line, and the pop-off line as required. Remote filler pipes, spit valves, and the pop-off valves are only required when a fuel tank is installed in the trunk of a passenger car or inside an enclosed compartment in a truck or van. If a topper is installed over the bed on a pickup truck with a

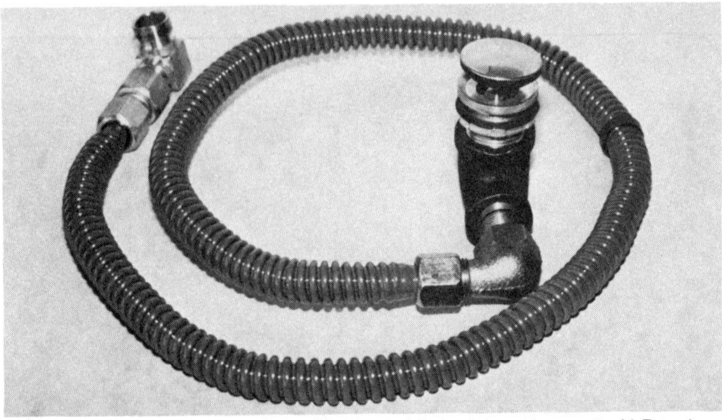

Fig. 8-9. Pop-off pressure relief valve assembly for internal-mounted LP tanks.

bed-mounted tank, a remote filler pipe, a spit valve, and a pop-off valve must be installed.

Install the fuel line from the fuel tank to the engine compartment using the recommended braided pressure hose and fittings. A hydrostatic relief valve must be installed in the fuel line between the fuel tank and fuelock. The relief valve must be located away from any heat sources and protected from grease and dirt.

Put a couple of gallons of propane into the fuel tank and open the manual shut-off valve for a few seconds to clear the fuel line of any dirt or debris. Do this outdoors and away from any open source of flame. Then shut the valve.

Connect the loose end of the fuel line to the fuelock. Open the manual shut-off valve and check the lines for leaks with an electronic leak detector or soapy water. If soapy water is used, no bubbles should appear around any of the fuel line connections and the system should hold pressure. Leave the manual shut-off valve open. The fuel system installation is now complete.

This step is optional. Remove the heat rise valve from the exhaust manifold or wire it in the open position. This modification is not recommended on applications where gasoline will be used frequently because the engine will hesitate and stumble when cold when first started on gasoline.

The engine should now be ready to start. Make a final check of all systems and connections. Shut off the propane system using the dash controls.

Start the engine on gasoline and allow it to reach operating temperature (Fig. 8-10). Then switch it over to propane following the basic procedure described at the end of this chapter.

Fig. 8-10. The engine should be allowed to reach operating temperature before any fuel mixer adjustments are made.

Make any adjustments to the ignition timing and set the idle speed and mixture settings on the fuel mixer (Figs. 8-11 through 8-13). The vehicle should then be run on a chassis dynamometer to make the fuel mixer power adjustments with the engine loaded at road speed.

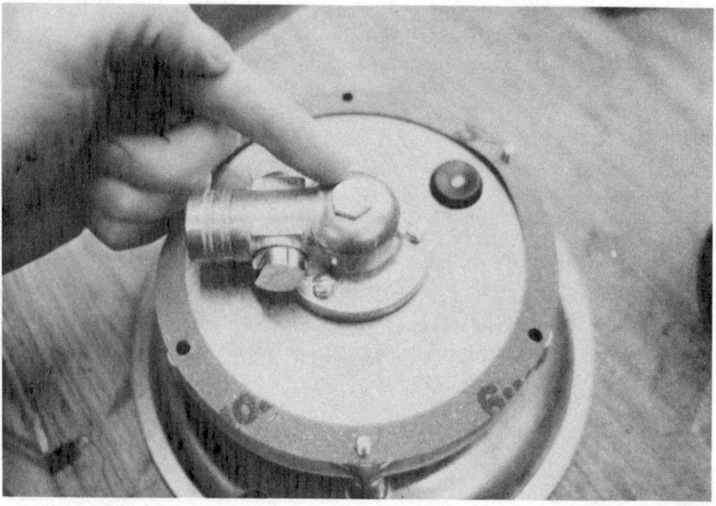

Fig. 8-11. The idle-mixture adjustment on the Impco mixer is located under the hex screw in the center of the unit. Remove the hex screw and adjust the idle mix with a screwdriver.

Fig. 8-12. The power adjustment on an Impco mixer. This adjustment is best made with an engine under load on a chassis dynamometer.

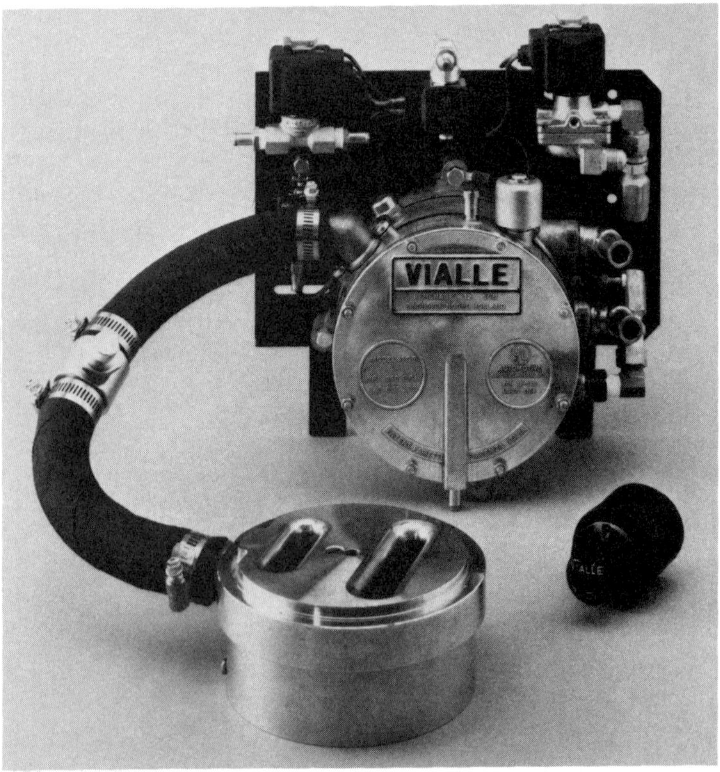

Fig. 8-13. A Vialle dual-slot propane gas mixer and converter for Chevrolet vans and trucks equipped with 305 and 350 V-8 engines. The mixer is installed on top of the existing 4-barrel carburetor to make a dual-fuel conversion.

With the adjustments complete, fill the fuel tank with propane and recheck the system for leaks. Then road test the vehicle. Switch back and forth from one fuel to the other to make sure everything works properly.

The dual fuel conversion is now complete. The only other thing that must be done is to educate any potential drivers about switching from one fuel to the other. The best way to do this is to take them for a ride and demonstrate the proper procedures.

SWITCHING FROM PROPANE TO GASOLINE

Simply pull the control lever all the way out. This shuts off the propane and opens the gasoline fuel line solenoid. It usually takes a couple of seconds for the carburetor bowl to fill with gasoline. Expect a slight hesitation during the transition.

SWITCHING FROM GASOLINE TO PROPANE

Push the control lever only half way in. This shuts off the flow of gasoline to the carburetor. Wait a few seconds until the engine begins to lose power; this indicates that all the gasoline has been used from the carburetor. Then push the control lever all the way in to turn on the propane.

Shoving the control lever in without waiting for the carburetor to run dry can flood the engine with fuel. This could cause it to stall. *Note:* The control cable positions might be exactly the opposite on some manufacturer's equipment. With some controls, *in* is for gasoline and *out* is for propane. Follow whatever is printed on the dash-mounted control panel.

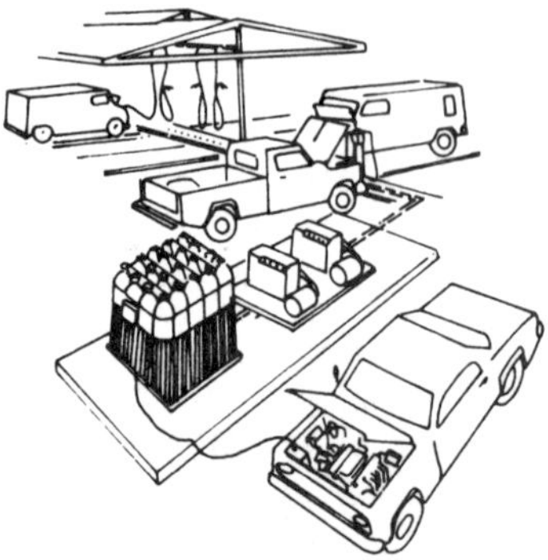

Chapter 9

Safety

Safety can't be over-emphasized when you are installing propane conversion equipment. A properly installed propane system is safer than a gasoline system. See Figs. 9-1 through 9-4. But an installation done improperly either as the result of carelessness or ignorance creates a potential for disaster. It's very important, therefore, that anyone who does conversions be familiar with all the safety considerations outlined in the National Fire Protection Associations's Pamphlet #58, local and state regulations, and the voluntary certification program sponsored by the National LP-Gas Association. A safety inspection checklist, a reference summary of the NFPA standards, and a description of the NLPGA's certification program follow.

The Tank

______ Name Plate: Mfg. name, date of mfg, size in lb. or wc., type of service, tare weight, surface area, design pressure.
______ Relief valve (internal type)
______ Liquid valve
______ Filler valve
______ Fixed liquid level gauge
______ Magnetic float gauge (could have a remote sender)
______ Container location (sealed from passenger area) with as much road clearance as practicable, but not less than the minimum road clearance under maximum spring deflection
______ Securely fasten container so it will not jar loose, slip or rotate
______ Valves and appurtenances protected
______ Sized properly (200 gal. wc or less for passenger carrying vehicles)

Fig. 9-1. Propane conversions must conform to national and local safety standards. That means fuel tanks must be ASME rated. Converters and fuelocks must be UL-approved for propane. Substandard or "make do" parts should never be used.

______ If tank is in trunk, the trunk should be vented on the bottom to displace air and gas
______ Remotely filled-double back flow check must be placed in the tank and piped outside the vehicle with approved* hose
______ The relief valve pipeaway shall be within 45° of vertical so it will not impinge on the vehicle or adjacent vehicle
______ Relief value discharge lines shall be metallic and properly sized (not aluminum); flexible metal hose or tubing, if used, shall be able to withstand the relief valve discharge pressure
______ Trunk sealed (passenger vehicle) including radio speakers
______ Seal two-way radio if enclosed with tank

Hoses and Piping

______ Hoses in excess of 60″ must be stainless steel, wire braid, reinforced marked for LP-gas service (350 psig working pressure, 1,750 psig burst pressure)
______ Hydrostatic relief valve shall be rated between 400-500 psig and protected with rain cap
______ Piping installed and supported to minimize the possibility of damage

Engine Compartment

______ Vacuum lock off or solenoid approved*
______ Hose to vaporizer regulator approved* 350 psig/1,750 psig burst pressure
______ Carburetor approved* if pressurized

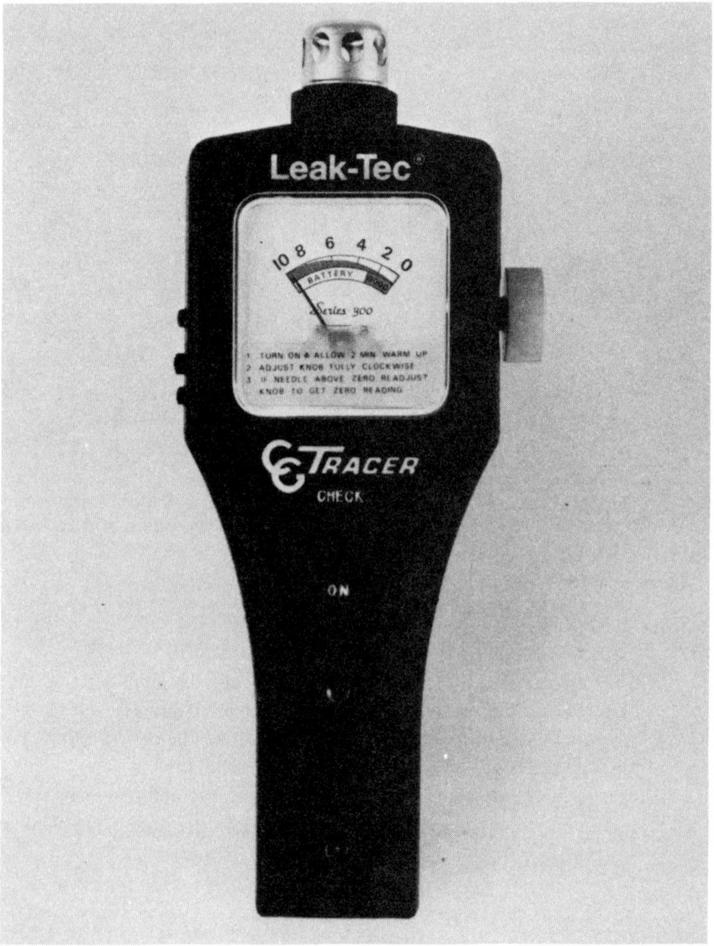

Fig. 9-2. An electronic sniffer for checking fuel line connections. This unit from American Gas & Chemical can detect concentrations of gas as low as 2 parts per million. Electronic sniffers can also be installed in the trunk to continually monitor for leaks.

_____ Vaporizer regulator approved* properly sized for engine
_____ Fuel pump removed and blocked if straight LP-gas powered

Check Valves, Fittings, Pipes, Hoses for Leaks with Soap Solution

*As defined in NFPA Pamphlet #58 acceptable to the authority having jurisdiction or listed by acceptable testing laboratory.

NFPA 58 Reference

I. CONTAINERS

2102. Containers shall be designed, fabricated, tested and marked (or stamped) in accordance with the regulations of the U.S. Department of Transportation (DOT), the Rules for the Construction of Unfired Pressure Vessels, Section VIII, Division 1, ASME Boiler and Pressure Vessel Code, or the API-ASME Code for Unfired Pressure Vessels for Petroleum Liquids and Gases applicable at the date of manufacture . . .

2154. ASME Containers shall be marked . . . and the markings specified shall be on a metal nameplate attached to the container so located as to remain visible after the container is installed.

(a) Service for which the container is designed . . .

(b) Name and address of container supplier or trade names of container.

(c) Water capacity of container in pounds or U.S. gallons.

(d) Design pressure in psig.

(e) The wording "This container shall not contain a product having a vapor pressure in excess of ______ psig at 100F." . . .

(f) Tare weight of container fitted for service for containers to be filled by weight.

(g) Outside surface area in square feet.

2202. Container appurtenances shall have a rated working pressure of at least 250 psig.

3810. (a) ASME containers shall be constructed for a minimum 250 psig design pressure.

(b) Containers installed in enclosed spaces on vehicles, (including recesses or cabinets . . . shall be constructed as follows: (1) DOT cylinder specification containers shall be designed and constructed for at least a 240 psig service pressure. (2) ASME

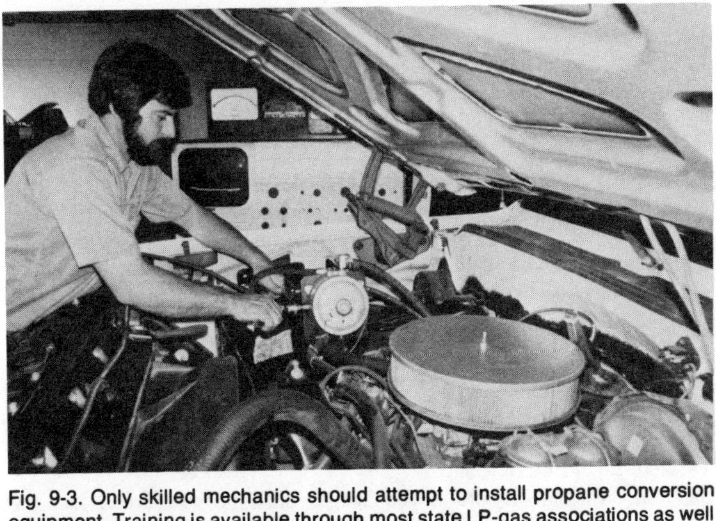

Fig. 9-3. Only skilled mechanics should attempt to install propane conversion equipment. Training is available through most state LP-gas associations as well as numerous equipment manufacturers and distributors.

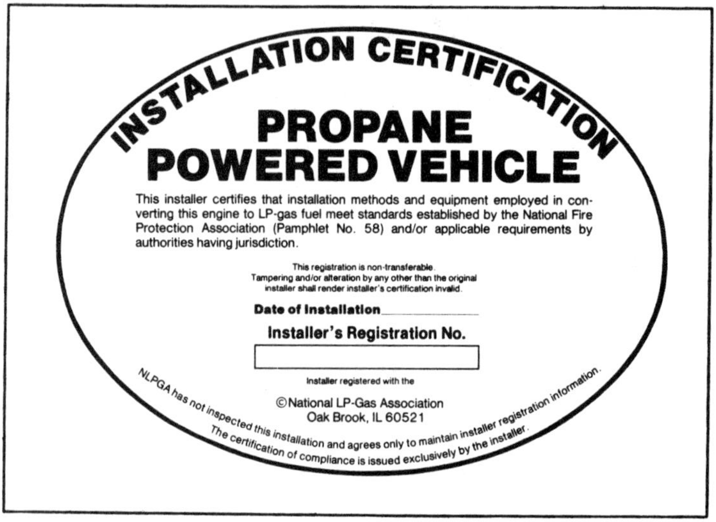

Fig. 9-4. This is what the NLPGA's certification decal looks like. It certifies that the propane conversion was installed properly and safely.

containers shall be constructed for at least a 312.5 psig design pressure. (3) Engine fuel containers shall be equipped with internal type spring loaded safety relief valves. Fusible plugs shall not be used.

(e) Individual LP-gas fuel containers used on passenger carrying vehicles shall not exceed 200 gallons water capacity.

II. VALVES AND GAUGES

Liquid Shutoff Valve

2222. (a) Manual shutoff valves shall be designed to provide positive closure under service conditions.

(Table 2-3) and internal excess flow check valve.

Fixed Liquid Level Gauge

2231. Every container constructed after December 31, 1965, designed to be filled on a volumetric basis, shall be equipped with a fixed liquid level gauge(s) to indicate the maximum filling level(s) for the service(s) in which the container is to be used . . . This may be accomplished either by using a dip tube of appropriate length, or by the position of the gauging device in the container . . .

2233. Gauging devices requiring bleeding of product to the atmosphere such as fixed liquid level, rotary tube, and slip gauges, shall be designed so that the bleed valve maximum opening to the atmosphere is not larger than a No. 54 drill size, unless equipped with excess-flow check valves.

Variable Liquid Level Gauges

2232. (a) Variable Liquid Level Gauges shall be so marked that the maximum liquid level, in inches or percent of capacity of container in which they are to be installed, is readily determinable. These markings shall indicate the maximum liquid level for propane, for 50/50 butane-propane mixtures, and for butane at liquid temperatures from 20°F to 130°F and in increments not greater than 20 degrees.

NOTE: Except for remote senders, magnetic float gauges have the same marking as variable gauges.

Filler Valve and Double Flow Check

2222. (c) Back flow check valves, which may be of spring-loaded or weight-loaded type with the in-line or swing operation, shall close when flow is either stopped or reversed. Both valves of double back flow check valves shall comply with this provision.

Relief Valves

2210. Containers shall be equipped with . . . safety relief devices which . . . shall be designed to relieve vapor.

2213. (b) Each safety relief valve shall be plainly and permanently marked with: (1) The pressure in psig at which the valve is set to discharge: (2) the actual discharge rate in cubic feet per minute of air at 60 F and 14.7 psia; and (3) the manufacturer's name and catalog number. Example: A safety relief valve is marked 250-4050 AIR. This indicates that the valve is set to start-to-discharge at 250 psig; and that its rate of discharge is 4050 cfm of air.

III. HOSES

Stainless Steel Wire Braid

3512. On engine fuel lines on vehicles where a hose in excess of 60″ in length is used, the hose shall have stainless steel wire braid reinforcement . . .

2350. Hose, hose connections and flexible connectors . . . shall be fabricated of materials resistant to the action of LP-gas both as liquid and vapor. If wire braid is used for reinforcement it shall be of a corrosion resistant material such as stainless steel.

2352. (a) Hose shall be designed for a minimum bursting pressure of 1,750 psig, (350 psi working pressure) and shall be marked with "LP-GAS" or "LPG" and with the working pressure in psig at not greater than 10 foot intervals.

(b) Hose assemblies, after the application of connections, shall have a design capability of withstanding a pressure of not less than 700 psig. If a test is made, such assemblies shall not be leak tested at pressures higher than the working pressure (350 psig minimum) of the hose.

Hose of Less Than 60″

2350. Hose, hose connections and flexible connectors . . . shall be fabricated of materials resistant to the action of LP-gas both as liquid and vapor. If wire braid is used for reinforcement it shall be of a corrosion resistant material such as stainless steel.

2352. (a) Hose shall be designed for a minimum bursting pressure of 1,750 psig, (350 psi working pressure) and shall be marked with "LP-GAS" or "LPG" and with the working pressure in psig at not greater than 10 foot intervals.

(b) Hose assemblies, after the application of connections, shall have a design capability of withstanding a pressure of not less than 700 psig. If a test is made, such assemblies shall be leak tested at pressures higher than the working pressure (350 psig minimum) of the hose.

IV. TUBING

2320. (c) Copper tubing

(1) Type K or L, ANSI H 23.1 "Seamless Copper Water Tube," (ASTM B88-75a).

(2) ANSI H23.5, ASTM B280-75a, "Seamless Copper Tube for Air Conditioning and Refrigeration Field Service."

V. HYDROSTATIC RELIEF VALVES

3816. (h) Hydrostatic relief valves . . . shall be installed in isolated sections of liquid piping

2360. Hydrostatic relief valves, designed to relieve the hydrostatic pressure which might develop in sections of liquid piping between closed shutoff valves shall have pressure settings not less than 400 psig or more than 500 psig . . .

VI. SOLENOIDS

3511. (b) An approved automatic shutoff valve shall be provided in the fuel system at some point ahead of the inlet of the gas regulator, designed to prevent the flow of fuel to the gas air mixer when the engine is not running. Atmospheric type regulators (zero governors) shall not be considered as automatic shutoff valves except for portable engines of 12 horsepower or less with magneto ignition and used exclusively outdoors.

VII. VACUUM LOCKOFF

3511. (b) An approved automatic shutoff valve shall be provided in the fuel system at some point ahead of the inlet of gas regulator, designed to prevent the flow of fuel to the gas air mixer when the engine is not running. Atmospheric type regulators (zero governors) shall not be considered as automatic shutoff valves except for portable engines of 12 horsepower or less with magneto ignition and used exclusively outdoors.

VIII. VAPORIZER - REGULATOR

3510. (b) Vaporizers shall be fabricated of materials suitable for LP-gas service and resistant to the action of LP-gas under service conditions . . . Such vaporizers shall be designed for engine fuel service and shall comply with the following:

(1) The vaporizer proper, any part of it, or any devices used with it, which may be subjected to container pressure, shall have a design pressure of at least 250 psig, and shall be plainly and permanently marked at a readily visible point as follows:

(a) With the design pressure of the fuel containing portion in psig.
(b) With the water capacity of the fuel containing portion in pounds.

(2) Vaporizers shall not be equipped with fusible plugs.

(3) Each vaporizer shall have a valve, or suitable plug, located at or near the lowest portion of the section occupied by the water or other heating liquid to permit substantial complete drainage. The engine cooling system drain or water hoses may serve this purpose, if effective.

(5) Engine exhaust gases may be used as direct source of heat to vaporize the fuel if the materials of construction of those parts of the vaporizer in contact with the exhaust gases are resistant to corrosion from gases, and if the vaporizer system is designed to prevent excessive pressures . . .

3511. (a) Approved automatic pressure reducing equipment, properly secured, shall be installed between the fuel supply container and the gas-air mixture to regulate the pressure of the fuel delivered to the gas-air mixture.

IX. RELIEF VALVE PIPEAWAY

3812. (d)(3)(b) The discharges from . . . container relief valves shall be piped to the outside of the vehicle. Relief valve discharge shall be directed upward within 45° of the vertical so that any gas released will not impinge on the vehicle, and so that the possibility of impingement on adjacent vehicles is minimized.

3814. (a) Container safety relief valves shall be located and installed as follows:

(2) Relief discharge outlets on containers installed on the outside of the vehicle shall be located:
(a) Outside of enclosed spaces, at least 3 feet horizontally away from any opening into vehicle below the level of such discharge, and as far as practicable from sources of ignition.
(b) In such a manner as to minimize the possibility of impingement of escaping gas upon a container, vehicle parts, or on other vehicles in adjacent lines of traffic. For engine fuel containers, the relief valve discharge shall be vented upward within 45° of the vertical to accomplish this.

(3) Safety relief valve discharge lines, if used shall be metallic (other than aluminum) and shall be sized, located and secured so as to permit sufficient safety relief valve discharge capacity. Flexible metal hose or tubing used shall be able to withstand the pressure from the relief valve vapor discharge when the relief valve is in full open position.

(4) On vehicles used outdoors or in industrial locations, means shall be provided (such as loose fitting caps) to minimize the possibility of the entrance of water or dirt into either the relief valve or

its discharge piping. The protecting means shall remain in place except when the relief valve operates. In this event, it shall permit the relief valve to operate at sufficient capacity.

X. MOUNTING OF CONTAINERS

3812. Containers shall be securely mounted on the vehicle, or within the enclosing recess or cabinet, and located and installed so as to minimize the possibility of damage to containers, their appurtenances or contents as follows:

(a) Containers shall be installed with as much road clearance as practicable but not less than the minimum road clearance of the vehicle under maximum spring deflection. This clearance shall be measured to the bottom of the container, or to the lowest fitting, support or attachment on the container or container housing whichever is lower.

(b) Fuel containers shall be securely mounted to prevent jarring loose and slipping or rotating, and the fastenings shall be designed and constructed to withstand without permanent visible deformation static loading in any direction equal to four times the weight of the container filled with fuel. When containers are mounted within a vehicle housing, the securing of the housing to the vehicle shall comply with this provision. Any hoods, domes or removable portions of the housing or cabinet shall be provided with means to keep them firmly in place in transit . . .

(c) All container valves, appurtenances and connections shall be adequately protected to prevent damage due to accidental contacts with stationary objects from loose objects, stones, mud, or ice, thrown up from the ground or floor, and from damage due to overturn or similar vehicular accident. In the case of permanently mounted containers, this provision may be met by the location on the vehicle, with parts of the vehicle furnishing the protection . . . Such weather protection as may be necessary to insure safe operation shall be provided for containers and systems mounted on the outside of the vehicle.

(d) Engine fuel containers on passenger carrying vehicles may be installed in the luggage compartment (trunk) of passenger cars and taxicabs, or in spaces completely sealed off from the interior of station wagons, vans or other similar vehicles as follows:

(1) Containers shall be located in a place and in a manner to minimize the possibility of damage to the container and its fittings. Containers located in the rear of the vehicles, when protected by substantial bumpers, shall be considered in the conformance with this requirement. In case the fuel container must be installed near the engine or exhaust system, it shall be shielded against direct heating.

(2) The passenger carrying compartment, and any space containing radio equipment, shall be isolated from the space in which the

container and container fittings are located by a seal to prevent direct seepage of gas to these spaces by venting this sealed off container space to the outside of the vehicle so that:

(a) Any LP-gas released within it will be readily dissipated to the outside, and

(b) The seal will not be damaged due to the sudden increases in air pressure which may occur when the luggage compartment or vehicle doors are closed quickly.

(3) Permanently installed fuel containers on passenger carrying vehicles shall be installed and fitted so that no gas from fueling and gauging operations, or from relief valves, can be released inside the passenger or luggage compartments, or within any space containing radio equipment, as follows:

(a) Remote (outside) filling connections shall be used . . .

XI. INSTALLATION OF PIPING

3816. (d) The piping system shall be designed, installed, supported and secured in such a manner as to minimize the possibility of damage due to vibration, strains or wear, and to preclude any working loose while in transit.

(e) Piping shall be installed in a protected location, and if outside under the vehicle and below any insulation or false bottom, fastenings and protection shall be such as to prevent abrasion or damage due to vibration. At points where piping passes through structural members or floors, a rubber grommet or equivalent shall be installed to prevent chafing.

(f) Gas piping shall be installed to enter the vehicle through the floors directly beneath, or adjacent to, the appliance served. If a branch line is required the tee connection shall be in the main gas line under the floor and outside the vehicle.

(g) Exposed parts of the piping system shall either be of corrosion-resistant material or adequately protected against exterior corrosion.

(i) Piping systems, including hose, shall be tested and proven free of leaks . . .

(j) There shall be no fuel connections between a tractor and trailer or other vehicle units.

NLPGA SAFETY CERTIFICATION PROGRAM

A voluntary self-certification program has been developed by the National LP-Gas Association to provide a way of identifying and selecting installers who are fully qualified and who follow all industry accepted safety standards. To be certified, a participant in the program must perform propane conversions in accordance with industry standards and use only system components that are fully approved. After the applicant has completed a form certifying compliance with these requirements, the installer is added to the national registry maintained by the NLPGA. The installer is then permitted to affix a certification decal to each vehicle he converts.

As use of the decals becomes widespread, the NLPGA will encourage LP-gas dealers to sell propane only to those vehicles that have the approved decal. The idea is to provide an incentive for conversion centers to participate in the program.

QUESTIONS AND ANSWERS

The following questions and answers should help explain the certification program.

Why was the NLPGA's self-certification program developed?

There are three basic reasons. First, and most important, the NLPGA wants to help prevent unsafe installations of propane fuel systems by unqualified installers. Second, the NLPGA wants to prevent the industry's outstanding safety record from being unjustly tarnished. In addition, the NLPGA wants to provide its members with a tangible, voluntary way to demonstrate their commitment to safety. The self-certification program meets all of these criteria.

Just how serious is the problem of unsafe installations?

That really depends on the definition of "serious." Like gasoline, propane is combustible; an improperly installed system is a safety hazard. The actual number of accidents each year involving propane-powered vehicles is very low. Many governmental agencies and independent testing organizations report that a properly installed propane fuel system is safer than a gasoline system. The excellent safety record of propane during the last 40 years is strong evidence that the vast majority of conversions to propane have been done properly by competent personnel. This isn't surprising when you realize that, until the last year or two, almost all propane conversions were done by LP-gas dealers who were fully aware of the potential hazards—and the appropriate safety measures—of propane carburetion. If you equate the term "serious problem" with thousands of unsafe vehicles on American highways, the problem today is not serious.

Then why the concern?

Every unsafe installation is an unnecessary safety hazard. The

dents can be prevented by having conversions done only by in-
NLPGA wants to make sure that fleet operators realize that acci-
stallers trained in propane carburetion. This will not only help save lives, it will also help prevent propane from getting a bad reputation with the public.

How's that?

Although accidents involving propane-powered vehicles are rare, those that do occur often take on disproportionate significance with the public, the media, and with legislators. Everyone is familiar with gasoline and its hazards, but propane is the "new alternative" and as such it's automatically suspect. When an accident happens, people become alarmed and call for special control of propane—despite its excellent safety record.

The NLPGA is convinced that propane can play a significant role in helping deal with the nation's energy problems, and unsafe or improper installations can jeopardize that role.

If accidents are so few, why is the NLPGA concerned that propane's role as an alternative engine fuel could be jeopardized?

Because as the number of propane conversions increases—it's grown from about 35,000 in 1978 to 350,000 in 1981—so does the potential for unsafe installations. More unsafe installations can translate into accidents. Something must be done now before the problem has a chance to grow. Additionally, unskilled installers are likely to lack proper tools and equipment. Even if the job is nominally correct, engine performance is likely to be poor. This will give the fuel—not the installer—a black eye.

Does an increase in the number of conversions have to result in a corresponding increase in unsafe conversions?

Not if the conversions are all done by installers who know what they're doing. LP-gas dealers have been just about the only qualified installers available, and the rapid growth in demand for conversions has been difficult for them to handle. It's not uncommon for an LP-gas dealer, who installs propane fuel systems, to have a lengthy waiting list of fleet customers and individuals. Cost is often a factor too. Some people simply get tired of waiting, and seek other alternatives—quicker or merely "cheaper."

Such as?

Some "do-it-yourselfers" have tried installing their own system—with tragic results. Some fleet operators have had mechanics without special training install systems using "kits" and "manuals" that fall short of industry-accepted safety standards.

There have also been instances of "propane carburetion centers" set up—at least for a while—by self-styled experts whose only qualifications are that they know a little about auto servicing and are looking for a way to make some fast money.

If the converted vehicle does not perform satisfactorily, at best, a customer blames only propane (nothing else) and complains to anyone and everyone who will listen. Part of the problem is that propane fuel systems are simple, and installation procedures do not appear to be complex. But even an experienced, skilled mechanic can overlook crucial safety procedures if he hasn't been adequately trained in them. When he takes on a propane conversion, he's gambling that everything will go just fine. But if he makes a mistake, it will be the driver of the vehicle who loses.

How will the certification program help solve the problem of unqualified installers?

The NLPGA thinks the program will help in two ways. First, it will help prospective conversion customers understand the hazards of an unsafe conversion. Second, it will provide them with a way to differentiate those who are prepared to certify in writing that they comply with accepted standards from those who aren't. The NLPGA is confident that, once buyers understand what they're dealing with, they'll take their business to registered installers.

As further "encouragement" for vehicle operators to deal only with registered installers, the NLPGA will attempt to persuade retailers to dispense propane only into vehicles that carry the self-certification program decal. A customer who has a conversion done by an unregistered installer might run the risk of not being able to buy fuel. Of course, the NLPGA can't compel dealers to comply, but it's in their best interest to do so.

How does the program work?

It's really quite simple. The installer must complete an application in which he certifies that he will comply with three requirements:

■ To perform all installations in accordance with the provisions of National Fire Protection Association Standard 58 or other applicable state, local, or federal regulations.

■ To use only components and equipment that meet the specifications of NFPA Standard 58 or other applicable regulations (that means many elements must be UL listed).

■ To carry the minimum level of product liability insurance required by applicable jurisdictions.

In return, he is entitled to identify his operation as an installation facility registered with NLPGA and to affix certification decals bearing his official registration number to each vehicle he converts. The decal not only provides the customer with verification that the job was done safely, it might also eventually be needed to get propane fuel. Should it ever be necessary to identify where an installation was done, the installer's identifying number on the decal is registered with NLPGA.

What about vehicles that have already been converted?

The NLPGA anticipates that many participants in the self-certification program will establish programs for "retro-decalling." Some NLPGA members plan to notify all their conversion customers to come in for a free inspection and decal. Others will inspect for a fee and certify the safety of installations done elsewhere, provided the vehicle measures up to safety standards. As the program builds momentum, anyone operating a properly-converted, safe vehicle will be able to get a certification decal.

Does participating in the certification program increase liability for an installer?

Yes, and the NLPGA stressed that fact to registrants. There has always been a certain inherent liability for any installation, but the NLPGA's lawyers say that the liability is increased for someone who states that specific safety procedures are followed, but does not do so. The initial participants will be companies or individuals with a firm commitment to safety. They will be people who would do the job right with or without a certification program.

Ultimately, all installers who want part of the growing conversion business will participate in the program because customers won't patronize nonparticipants. The added liability they assume will be a powerful motivator to make sure that they live up to their safety commitment.

Liability can be a strong motivator to do the job properly, but it is still a voluntary program. Are any other measures being considered?

The NLPGA is evaluating a number of possible approaches. At the state level a few very rigorous, mandatory programs are being established. NLPGA members view the self-certification program as step one of a vitally needed process. If stronger measures are required to assure safety, they'll be taken.

What's being done to increase the number of qualified installers?

Today most conversions are done within the LP-gas industry

by dealers through conversion centers or by equipment distributors. This will change because there is a definite need for more qualified installers to keep pace with the demand for conversions. The goal is not to create a monopoly of the conversion business, but to make sure that everyone involved is doing the job safely and properly. Auto/truck repair facilities of all types could participate in the certification program after their personnel have been trained. Some LP-gas companies already offer such training to service outlets in their area, and this type of program will probably grow. Some very large propane companies are also actively seeking out reputable well-equipped automotive service centers to do conversions for them.

Chapter 10

Tuning and Troubleshooting

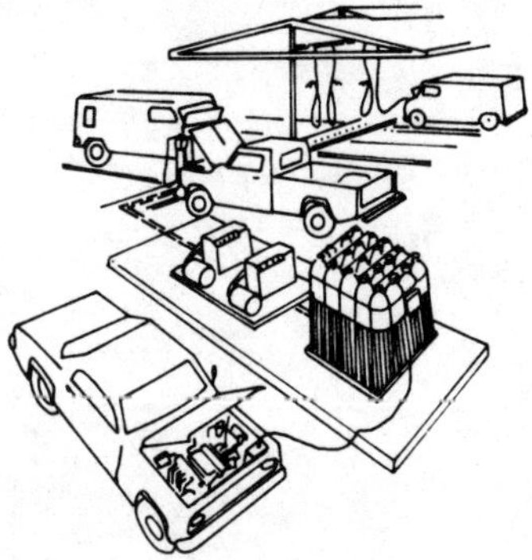

The four basic adjustments to make when tuning a propane-powered engine are the ignition timing, idle speed, idle mixture, and power mixture. For the engine to perform properly and to deliver good fuel mileage, these four items must be set accurately.

IGNITION TIMING

To set the ignition timing, connect a timing light to the number one spark plug wire. Disconnect the vacuum hose from the distributor diaphragm (on units equipped with vacuum advance), and plug the hose with a small screw or piece of dowel. A golf tee also makes a good temporary vacuum hose plug. Loosen the distributor hold-down nut only enough so that the distributor can be turned by hand—but don't turn it yet. Start the engine and shine the timing light on the front pulley or flywheel timing marks. Now turn the distributor until the desired timing is achieved.

On straight-propane conversions, the initial timing is typically advanced 3 to 6 degrees over the stock setting. On dual fuel conversions, the timing is advanced only 2 or 3 degrees (Fig. 10-1).

For high altitude applications, some compensation must also be made for the thinner air. Generally, the timing should be advanced 4 degrees for every 5000 feet of elevation. If the altitude compensation isn't made, count on a loss of horsepower and possibly shorter exhaust-valve life.

Once the initial timing is set, tighten down the distributor hold-down nut and recheck the timing to see that it hasn't changed.

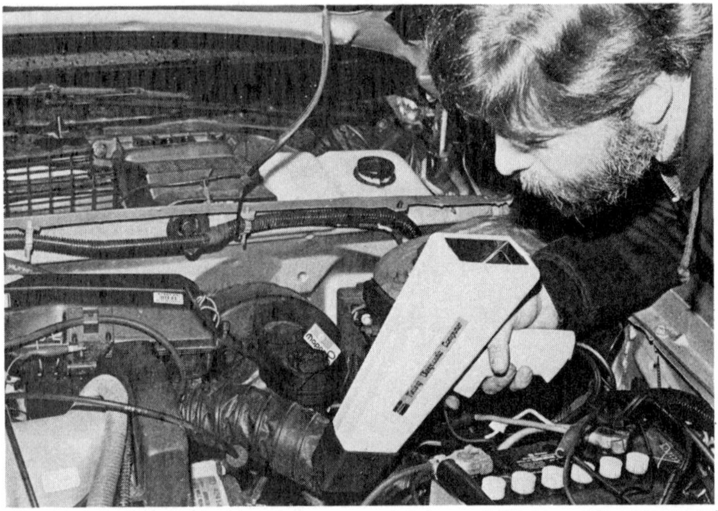

Fig. 10-1. Setting the ignition timing accurately is crucial to top performance and economy. Some timing lights will also show degrees of total advance. This is very helpful with modified advance curves.

Now rev the engine up and watch the timing marks. If the centrifugal advance mechanism is working properly, the timing will advance. Reconnect the vacuum hose to the advance diaphragm and watch the timing marks again. If the vacuum hose is connected directly to the manifold, you should see the timing advance. On some engines, vacuum advance doesn't begin until the carburetor throttle plates open slightly. This is called ***ported vacuum.*** On many late-model engines with emission controls, there are vacuum delay valves and switches. Refer to the vehicle service manual for the recommended procedure to check vacuum advance on such vehicles.

Total spark advance should also be checked because over-advanced timing can cause detonation. If not prevented, heavy detonation will eventually cause serious engine damage. Therefore, it is important that the total timing advance be checked using one of the following methods.

■ The most accurate method is to remove the distributor from the engine and recalibrate it on a distributor machine (Fig. 10-2).

■ The second way to measure total spark advance is to use a timing light or engine analyzer that reads total advance electronically and displays the results on a meter or screen.

■ Another way is to take a pair of dividers and mark off additional timing marks on the pulley or flywheel using the existing timing marks as a guide (Fig. 10-3). In other words, if a half inch

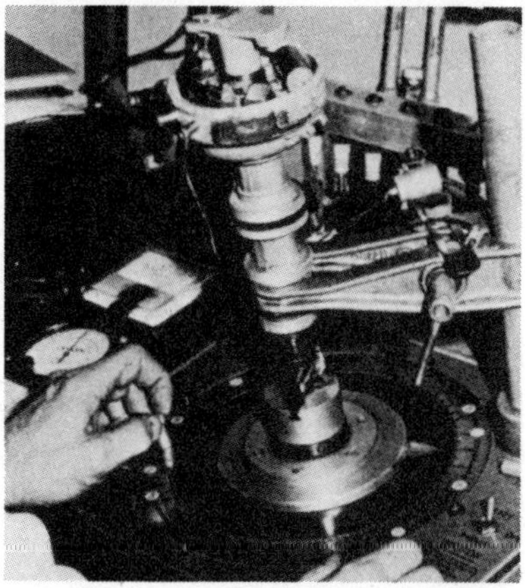

Fig. 10-2. The advance curve of the distributor can be recalibrated on a distributor machine.

equals 10 degrees at the pulley, marking off another inch and a half would indicate the 40 degree BTDC timing point. The maximum total advance should be marked on the pulley of flywheel with a notch or paint. This paint should not be exceeded when the engine is run at maximum rpm.

For heavy-duty truck engines that typically operate at low rpm, maximum total advance should not exceed about 28 to 32 degrees. On higher-reving automobile engines, total advance should not exceed 38 to 40 degrees.

All too often a propane conversion turns sour because the mechanic simply set the initial timing and never checked the total advance. He assumed it would be all right when it was over-advanced. Don't make the same mistake.

One way to limit the amount of total advance is to restrict the travel of the flyweights in the advance mechanism. The Gann

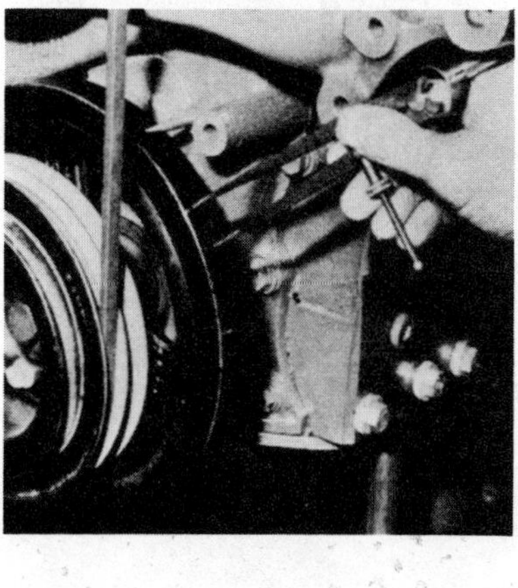

Fig. 10-3. Using dividers to mark off additional timing marks on the crankshaft damper. This allows the mechanic to check total timing advance with a timing light.

Products Company of South Gate, California manufactures a small steel plate that fits inside the distributor (Fig. 10-4). The plates are available in increments of 5, 7, 9, and 11 degrees. It's important to note that 1 degree of distributor advance translates into 2 degrees of crankshaft advance. This is because the distributor turns at half the speed of the crankshaft. So if a Gann plate that allows 9 degrees of distributor advance is used, the engine's total advance will be 18 degrees of distributor advance, plus initial timing, plus vacuum advance. In many straight propane conversions, the vacuum advance is simply disconnected. This simplifies the job of timing and usually improves performance. Total advance is then controlled by centrifugal advance in the distributor and initial advance set at the pulley or flywheel.

IDLE SPEED

The idle speed on a straight conversion or a dual-fuel conversion should be set so that the engine idles at approximately the same rpm as before. Refer to the vehicle's service manual or a tune-up specification guide for the recommended rpm setting. For most engines, 600 to 800 rpm is the recommended setting.

Higher idle speeds are sometimes desirable for vehicles with standard stick-shift transmissions. The higher idle speed helps to reduce the chance of stalling when engaging the clutch. But for vehicles with automatic transmissions, the recommended setting should be used to keep the vehicle from creeping when it's in gear. Too high an idle speed can lead to accelerated automatic transmission and brake wear.

To adjust the idle speed, connect a tachometer to the ignition system and start the engine. Turn the adjusting screw on the propane carburetor or mixer until the desired rpm is achieved. Once

Fig. 10-4. A Gann centrifugal advance limited plate installed in a distributor. The contact point mounting plate has been removed to reveal the advance mechanism.

Fig. 10-5. Adjusting the idle mixture on a dual fuel mixer. Shown is an Impco unit.

set, the idle speed requires no further adjustment unless ignition timing changes are made.

IDLE MIXTURE

The idle mixture controls the air/fuel ratio when the engine is idling and at low speeds. The idle mixture adjustment is changed by turning a screw on the propane carburetor or mixer (Figs. 10-5 and 10-6).

To set the idle mixture adjustment, make sure the engine is at normal operating temperature. Next, use the *lean drop* method to find the smoothest idle. Turn the idle mixture screw in until the idle speed begins to drop. Then back it out slowly until the smoothest idle is achieved. Readjust the idle speed as required to maintain the recommended setting.

If a CO exhaust analyzer is used to check the adjustment, it should show about 1 percent carbon monoxide with the engine idling. A leaner mixture (lower CO reading) can cause a loss of power and possibly misfiring. A richer mixture (higher CO reading) will not get as good fuel economy and it creates more pollutants.

POWER MIXTURE

The power mixture screw is used to adjust the air/fuel ratio for full throttle operation. Light load or cruise mixtures are controlled by the shape of the gas metering valve and are not adjustable.

The power adjustment must be set with the vehicle operating at road speed under full load. This is best done on a chassis dynamometer using a CO exhaust analyzer. The power adjustment depends on the type of vehicle and how it will be used. A heavy-duty truck operating under continuous full-load conditions should be adjusted no leaner than 1.5 to 2.0 percent carbon monoxide. A light duty vehicle can operate safely as lean as 1.0 or 1.5 CO because it will not be running flat out for long periods of time.

The thing to keep in mind is that leaner mixtures (lower CO reading) get better fuel economy and richer mixtures (higher CO reading) generate more power. Extremely lean mixtures can result in misfiring because the added air raises the voltage requirements of the spark plugs. If the spark can't bridge the plug gap, the mixture won't be ignited.

Overly rich mixtures can elevate exhaust temperatures to the point where exhaust valve life is shortened. Too rich mixtures can also foul the spark plugs, waste fuel, and increase pollutants. About 3 percent CO is considered to be the maximum richness for optimum performance. Horsepower will begin to drop as the power mixture is richened beyond this point. For practical purposes, there is not much of an increase in power from 2 percent CO to 3 percent CO.

TROUBLESHOOTING

Performance problems in propane-powered vehicles can be caused by equipment installation errors, misadjustments, or normal wear and tear. One of the most common equipment installation errors is mounting the converter upside down so the vapor hose outlet is at the top. This causes oil to collect inside the converter and this can deteriorate the pressure diaphragms. The vapor hose outlet should be located at the bottom or underside when installed properly.

Another mistake is mounting the converter, the fuelock or the fuel line too close to the exhaust manifold or exhaust pipe. The heat can cause the fuel to "boil" inside the fuel system. This can create dangerous pressures and possibly a vapor-lock condition.

Sometimes the converter hot water lines are not connected properly and water doesn't flow through the unit. Without the added heat, the converter can freeze up and stop the flow of fuel.

On dual-fuel conversions, a poor seal between the gasoline carburetor adapter and the fuel mixer will allow an air leak. This upsets the air/fuel mixture, and it can create idle problems and misfiring.

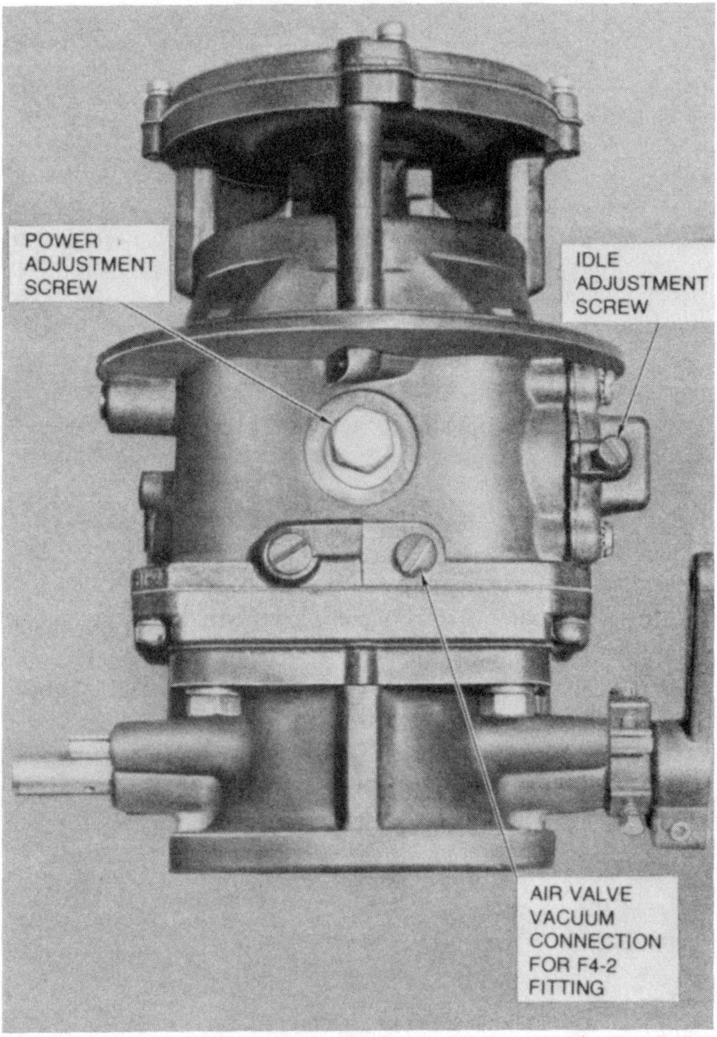

Fig. 10-6. Idle and power adjustment points on an Impco CA425 propane carburetor.

Misadjustments of the carburetor or ignition timing are usually the result of sloppiness or "trial-and-error" tuning techniques. There's no substitute for proper tune-up equipment. Tuning by ear can't begin to match the accuracy of a timing light, CO analyzer, or chassis dynamometer. Misadjusted timing and air/fuel mixtures can cause a variety of problems.

As for normal wear and tear, a propane-fuel system is far less prone to failure than a gasoline-fuel system. Failures occasionally do happen. The two most vulnerable components in the fuel system are the *carburetor air valve diaphragm* (Fig. 10-7) and the *converter primary valve seat* (Fig. 10-8). Contaminants in commercially available propane can sometimes cause the converter primary valve seat to fail. As for the carburetor air valve diaphragm, the synthetic rubber can be damaged by engine backfiring or mishandling during installation. The synthetic rubber usually lasts 50,000 to 100,000 miles, but extremely cold weather can cause the rubber to crack. Not driving the vehicle for a length of time could allow the rubber diaphragm to dry out and harden. The newer silicone-based synthetic diaphragms will not dry out with disuse.

The filter in the fuelock normally lasts 100,000 miles. Sometimes after the initial installation, the filter can be clogged by rust particles from inside the fuel tank. Rust is sometimes formed when the tanks are hydrostatically tested at the factory. If all the moisture isn't removed after testing, rust can form. A clogged filter will cause a pressure drop across the filter assembly. This will allow the fuel to vaporize inside the fuelock-filter, and cause frost to form on the outside.

Fig. 10-7. About the only thing that can go wrong with a propane carburetor is the air valve diaphragm. A cracked or torn diaphragm can be easily replaced in a few minutes, if necessary, with the carburetor on or off the engine.

Fig. 10-8. A faulty converter can usually be repaired without having to remove the unit from the car.

QUICK CHECK OF THE FUEL SYSTEM

One way to quick check the fuel system for malfunctions is to do the following.

■ With the air cleaner and carburetor air-gas valve and cover removed, the converter and fuelock can be checked for internal leaks by opening the manual fuel shut-off valve at the fuel tank. If the fuelock and converter are operating properly, no fuel will leak through the open carburetor gas valve.

■ Press the primer button on the converter to open the gas regulator valve inside. A small amount of propane should pass through the jet as the system is emptied back to the fuelock. If the fuelock is sealing the fuel line properly, the gas flow will cease as soon as the fuel between the fuelock and converter is used up.

■ Next remove the vacuum hose to the vacuum fuelock and apply a small amount of vacuum with a hand pump. On electric fuelocks, use a jump wire to open the solenoid. With the primer button depressed on the converter, fuel should flow immediately; it should stop as soon as the vacuum or jump wire is removed.

PROBLEM DIAGNOSIS

The following complaints and symptoms, and their possible causes, will help in troubleshooting fuel system and performance problems.

Stalled Engine

Out of fuel. Check the fuel level. When a propane-powered vehicle runs out of fuel, it doesn't sputter to a sudden stop as with gasoline. It will lose power, but will continue to run for awhile on the tank vapors.

Fuel not reaching engine. Frost on fuel lines or the fuelock is an indication of a blockage. Frost on the converter means hot water from the engine is not circulating through the unit.

No spark at spark plugs. Check for an ignition system problem such as a grounded coil, arcing around the coil high tension lead or rotor, or grounded or worn contact points.

Engine flooded (dual-fuel conversions only). Usually the result of not following the proper procedure when switching from gasoline to propane. Can also be caused by a leaky gasoline shut-off solenoid. Check for the presence of gasoline in the carburetor. There should be none when running on propane.

Engine Won't Start

Out of fuel.

Manual fuel shut-off valve at fuel tank closed. If valve was closed during installation or fuel system repair, someone might have forgot to open it. If closed, open it slowly so that the excess flow valve doesn't overload and snap shut.

Excess flow valve closed. Inside the manual shut-off valve is a second valve that prevents the excess flow of fuel in case of an accident. If the valve has shut itself for some reason, it should reset automatically when the excess flow ceases. To reset the valve manually, close the manual fuel shut-off valve and wait until a small click from inside the tank is heard. This usually takes about five seconds. Then slowly open the manual shut-off valve.

Plugged liquid fuel line. Rarely a problem after the initial conversion. If the line has been plugged with rust or debris from the fuel tank, close the manual shut-off valve at the tank. Disconnect the fuel line from the fuelock, and then open the manual shut-off valve for several quick bursts to blow out the fuel line. If the blockage cannot be removed, replace the fuel line.

Plugged fuelock filter. Filters normally go 100,000 miles or more before they require replacement. Rust or debris in a new fuel tank might cause plugging when first installed. Frost on the fuelock is a good clue to a plugged filter. Clean or replace the filter.

Insufficient tank pressure in extremely cold weather. If the temperature is −30 degrees Fahrenheit or colder, some type of tank heater might be necessary to generate sufficient pressure to get the vehicle started.

Faulty air-valve diaphragm in the propane carburetor or mixer. If the diaphragm has dried out and become hard or if it has cracks or pinholes, it must be replaced.

Fuelock not opening. Check for the presence of a strong vacuum-to-vacuum fuelock or a voltage-to-electrical fuelock. Vacuum leaks can be caused by loose or cracked vacuum hoses, a leaky intake manifold gasket, a leaky carburetor gasket, or air leaks between the gasoline carburetor and mixer adapter in dual-fuel conversions. No voltage or low voltage can be due to loose or corroded wire connections, a broken or grounded wire, a blown fuse, or a poorly grounded fuelock. If voltage is reaching an electrical fuelock, the solenoid might be defective. If vacuum is reaching a vacuum fuelock, the diaphragm might be ruptured.

Idle mixture too lean. Could be the result of a faulty adjustment or air leaks in the vapor hose connection between the converter and carburetor or mixer. Can also be caused by the carburetor air valve failing to open. This might be due to a damaged diaphragm, dirt in the carburetor, vacuum leaks around the carburetor, intake manifold or PCV valve, or the use of too large a carburetor for the application. A stiffened or faulty low-pressure regulator diaphragm in the converter can also restrict the flow of fuel to the carburetor.

Lack of priming.

Hard Starting

Idle mixture too lean. Could be caused by improper idle mixture adjustment, an air leak in the vapor hose between the converter and carburetor, a leak in the carburetor diaphragm, dirt in the carburetor, or vacuum leaks anywhere in the intake system. A stiffened or faulty low-pressure regulator diaphragm in the converter could also be the culprit.

Lack of priming. More of a problem at low temperatures.

Ignition timing over advanced or retarded. Check timing.

Weak spark. Check contact points and ignition coil.

Weak starter or low battery voltage. Check battery terminals and cable connections for corrosion or looseness.

Low engine compression. Due to worn valves or rings.

Engine Won't Keep Running

Clogged fuelock. Check for blockage and clean or replace filter.

Idle speed too low. Readjust for more rpm.

Idle mixture too rich. Readjust leaner. On dual-fuel conversions, richness can be caused by gasoline seeping past the fuel shut-off solenoid. Check for presence of gasoline in carburetor; none should be present when running on propane.

Idle mixture too lean. Could be caused by incorrect idle mixture adjustment, an air leak in the vapor hose between the converter and carburetor, a leak in the carburetor diaphragm, dirt in the carburetor, or vacuum leaks anywhere in the intake system. A stiffened or faulty low-pressure regulator diaphragm in the converter could also be at fault.

Lack of priming in cold weather.

Poor Acceleration

Excess flow valve inside the manual shut-off valve at the fuel tank closed. Reset manually.

Fuel mixture too rich or too lean.

Ignition timing over-advanced or retarded. Check timing.

Weak spark or worn spark plugs. Check contact points, ignition coil, spark plug cables, and spark plugs.

Balance line from air cleaner to converter too small (turbocharged and high-restriction air intake systems only). On such applications, the balance line should have a ⅜-inch inside diameter to avoid hesitation or backfire during acceleration.

Excessively warm air being drawn into air cleaner. On vans and some cabover installations, air inside the engine compartment can get very hot in warm weather. The engine's horsepower output drops about 1 percent for every 10-degree increase in intake air temperature. Because temperatures can soar to 180 degrees or more in tight engine compartments, the engine can lose up to 10 percent of its power. If overheated air seems to be a problem, install a duct to bring outside air to the air cleaner.

Engine Idles Rough

Idle speed too low. Readjust as required.

Incorrect idle mixture. Readjust for smoother idle.

Vacuum leaks in the intake system. Check all vacuum connections and hoses, carburetor and intake manifold gaskets, the PCV valve, and the mixer adapter on dual-fuel conversions.

Leaky engine valves. Check engine compression.

Weak spark or fouled spark plugs.

Over-retarded ignition timing. Check timing.

High Idle Speed

Vacuum leaks in intake system. Check all vacuum connections and hoses, carburetor and intake manifold gaskets, the PCV valve, and the mixer adapter on dual-fuel conversions.

Over-advanced ignition timing. Check timing. Centrifugal advance mechanism might be stuck in advanced position.

Incorrect idle speed adjustment. Readjust.

Sticking throttle linkage or weak throttle return spring.

Backfiring Through Carburetor

Incorrect power mixture. Readjust as required.

Defective engine intake valve. Check engine compression.

Cracked distributor cap, crossed ignition cables, or ignition crossfiring.

Lack of High Speed Power

Restricted fuel flow due to clogged fuelock filter, faulty converter, crimped fuel line, or too lean a power mixture adjustment.

Restricted exhaust system.

Over-advanced or retarded ignition timing.

High Fuel Consumption

Over-rich power mixture. Readjust to a leaner mixture.

Faulty converter or wrong low-pressure spring in converter for carburetor model.

Dirty air cleaner. Replace filter element.

Fuel system leaks. Use an electronic leak detector or soapy water solution to check all connections for leaks.

Retarded ignition timing. Check timing.

Low engine compression due to worn valves or rings. Check compression.

High rolling resistance. Check for dragging brakes or low tire pressure.

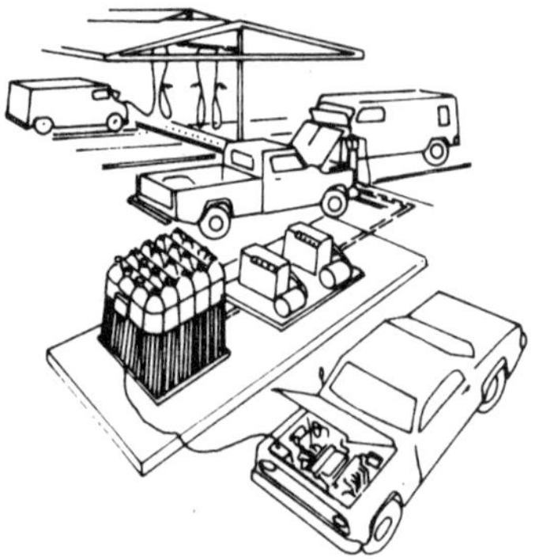

Chapter 11

Propane Conversion Businesses

A book about propane would not be complete without a chapter on the profit opportunities of the propane-conversion business. The information in this chapter is based on an article entitled "Propane Can Make You An Energy Czar" that appeared in the November, 1980 issue of *Entrepreneur Magazine*.

Even if you're not interested in starting a propane-conversion business, the following information will help you in dealing with those who do conversions. If you are a fleet operator, the description of the shop organization, equipment, and inventory will help you plan your fleet's conversion if you want to have your mechanics trained to do the job.

BIG OPPORTUNITIES

Across the country, entrepreneurs are netting up to 20 percent pre-tax profits on sales ranging as high as $60,000 a month by installing straight propane conversion systems in vans, trucks, and RV's or by putting dual-fuel (gasoline and propane) systems in automobiles (Fig. 11-1). A number of companies are now offering free or low-cost propane conversion training to potential operators of local centers. Franchises and dealer packages are also available.

NAMOCO, an Addison, Illinois-based distributor, offers a dealer investment package for about $2000. The price includes a complete dual-fuel conversion system with a 29-gallon fuel tank installed on the vehicle of your choice. Another complete dual-fuel

Fig. 11-1. The propane conversion business has netted some entrepreneurs a consistent 20 percent pre-tax profit on sales ranging as high as $60,000 a month.

conversion system includes a tank for your inventory, a comprehensive two-day training program at NAMOCO's training center, a copy of NFPA safety code pamphlet #58, and various sales aids and price sheets to get you started in the conversion business.

GROWING DEMAND

During the last gasoline crunch, manufacturers of propane careburetion equipment were backlogged six months or more. Although the situation has greatly improved thanks to increased production, another severe crunch could triple the conversion business.

In recent years, many large fleets have converted to propane as a means of reducing their operating costs. Interstate Brands, a California company that hauls fresh bakery goods, has converted its trucks to propane and reports its fuel mileage with propane is within two tenths of a mpg when compared to gasoline. Consolidated Freightways of Denver has converted its entire fleet to propane. The Los Angeles *Times* has converted its fleet and International Telegraph and Telephone (ITT) has done the same with many of its vehicles.

Crown-Zellerbach's Western Transportation Company converted its forklifts and found that propane not only solved excess exhaust emission problems, but that engine life of the lifts increased from 10,000 to 20,000 hours on gasoline up to 25,000 to 30,000 hours on propane. There are hundreds of examples of fleets and individuals who have cut their operating costs and extended maintenance intervals on vehicles that have been converted to propane.

With all the testimonials, it is no wonder that the demand for propane conversions has soared. Even original-equipment manufacturers are jumping on the propane bandwagon. Ford and International Harvester now offer propane options on many new truck models. Winnebago offers dual-fuel systems as a way of reducing the operating costs of its motorhomes.

The industry's biggest problem is finding enough competent conversion centers. There is a definite shortage because of the industry's sudden growth, and there are not enough trained mechanics to do conversions. Until the first gasoline crisis in 1973-74, propane conversions were most often done to commercial fleets of trucks, vans, and forklifts. The commercial operators were attracted by the pollution-free characteristics of the fuel and its lower operating costs. Once the gas crunch hit, consumer demand heated up and lines started forming at conversion centers.

Propane conversion centers are now attracting many small sized-fleets to medium-sized fleets. Included are trucking companies, taxis, municipal vehicles, garbage haulers, delivery operators as well as individuals who want their vehicles converted from using expensive gasoline. During 1981 in the United States, nearly 35,000 vehicles were converted to propane.

The surface has been barely scratched. There are 18 million fleets in the United States with five or more vehicles. And the total number of vehicles running on propane is still only about 2 percent of the total vehicle population.

The public's interest in switching to propane runs in cycles. When gasoline prices are high or when threats of a shortage appear, conversion centers can't handle the increase in business, and profits can run as high as 40 percent or more. When gasoline prices slack off and fuel supplies are readily available, the demand for conversions softens. In recent years, however, there has been a steady growth in the number of vehicles converted each year thanks to OPEC's continued stranglehold over world oil prices.

CONVERSION PROFITS

The typical conversion center changes from $800 to $1500 for a conversion. The conversion takes about eight hours with two men working on one vehicle. Cars usually take longer than trucks because on trucks there's more room to work. Cars also take longer because the trunk must be sealed and vented.

At Larry Osgood's Propac Industries conversion center in Fullerton, California, one conversion a day translates into $970 to as much as $2000 for some jobs. Forklifts usually run about $350 because they're so easy to convert. One conversion a day adds up to $20,000 to $40,000 a month with a gross profit between $10,000 and $15,000 per month.

According to NAMOCO's dealer profit program, two jobs per day at a selling price of $895 installed (not including tank or other extras) can net a dealer about $500 in profit per job, or about $1000 a day. You can see that a dealer who does two conversions a day on a regular basis can earn upwards of $250,000 a year. That does not include a profit the dealer can make on the tank and tank installation.

FLEET BUSINESS

There are an estimated 18 million fleets in the country (each with five or more vehicles). This represents a great business potential for the conversion center operator who aggressively seeks it out.

Major suppliers of propane, such as Petrolane, will sometimes convert large fleets at cost or slightly above just to get the fuel business on a contract basis. There are many other propane suppliers who don't want to get involved in the conversion business, and they will direct business to local conversion centers.

While it's hard for a beginner to compete with a huge entity such as Petrolane, the fleet market is virtually untapped nationwide. This is especially true for the small-sized fleets to medium-sized fleets the giants don't bother to go after. Propane conversion, according to insiders in the industry, is actually a relatively easy sell to fleet owners once they realize the tremendous savings propane offers. Lower fuel costs, lower maintenance costs, reduced downtime, and a ready supply of fuel are all selling pluses.

STARTING OUT

A primary concern when you are considering opening your own business is to locate in an area with as many potential customers as possible. For most businessmen this means a major metropolitan area. Make sure the area you have in mind has plentiful propane supplies available before you open shop. There's no use offering conversions if your customers can't buy fuel easily.

You could also become a propane dealer. This is an excellent way to generate a steady income and repeat business. By filling your customer's tanks, you can make from several cents up to 15 cents a gallon. A brand-new, 1000-gallon fuel storage tank and dispensing system would cost you around $5000. Or you could lease the equipment from a major supplier for as little as $25 a month in some areas. Some conversion center franchises include storage tanks as part of the package. You might want to have additional storage tanks on hand to sell or lease to local fleet operators who want the convenience of doing their own refueling.

The best way to check out the local propane supply situation in your area is to call the LP-gas companies listed in your telephone book. They should be able to quote you prices and give you information about fuel availability.

Unlike some businesses, you don't need high visibility with a conversion center. Propane conversions are not an impulse sales item. People think about it before they buy. After all, $800 to $1500 is a sizeable investment. Consider opening your conversion center in an industrial area where rents might be as low as 20 cents to 30 cents a square foot.

Another location that has good potential is the present or past site of a tune-up shop. Nearly half the start-up equipment you'll need is exactly the same as that which most tune-up shops use. Examples are hand tools, timing light, tachometer, engine analyzer, exhaust analyzer and a chassis dynamometer. If you happen to already own a tune-up shop or automotive repair business, the conversion business could be a natural and relatively inexpensive addition to what you're already doing. For those who are starting out cold, going into partnership with the owner of a tune-up shop or repair shop would save you much of the start-up costs (Fig. 11-2). It would also allow the entrepreneur to take advantage of the shop owner's existing automotive knowledge and skills.

Getting into propane work can also be a bonus for an existing tune-up service because most repair shops and service stations are unfamiliar with propane fuel systems. Although a propane tune-up is relatively simple once you know how to do it and have the right equipment, you can charge as much as $75 or more for your services because the job takes somewhat longer. The higher charge is justified because you're doing two tune-ups for the price of one on dual-fuel vehicles. First you tune the engine on gasoline, and then you tune it for using propane. Actually you're only tuning one

Fig. 11-2. Having well-trained personnel and the proper equipment is essential to running a successful conversion business.

ignition system and two carburetors; it really doesn't take that much longer to complete the job.

SHOP SETUP

The layout of your conversion center will vary depending on the site you select. About 2500 to 3000 square feet of inside space should be sufficient. It's better to have too much room than not enough. A typical shop might have four service bays; the drive-through format is the preferred layout. Two of the bays would be used for conversions and two for tune-ups and service. On one side of the shop should be the diagnostic equipment and tools required for tuning engines. Leave the other side relatively uncluttered so that the mechanics will have plenty of room in which to install parts.

The area where the conversions and tune-ups are performed should be well ventilated. Drive-through garages allow for better air circulation than single-door bays. Propane fumes are heavier than air and tend to settle to low areas. Make sure that there are no work pits or other potentially dangerous traps for the fumes to accumulate.

In addition to the service bay work area, you should have plenty of storage space for parts. You should have enough room to stockpile carburetors, adapters, hardware, and tanks. Storage space is crucial for two reasons. First, when you buy parts in volume there is a significant discount. You can buy your parts on an as-needed basis, but you'll be paying more than necessary. What's more, if another gasoline crunch develops, you might find yourself waiting for backordered parts. Second, distributors from which you buy parts are sensitive to demand, and prices during a crunch will usually go up sending your profits down.

You should also have a small office and a reception area. Be prepared to pickup and deliver your customers or vehicles because the installation time is too long for someone to wait in the shop.

The major pieces of shop equipment you need in order to do conversions are an exhaust analyzer, a chassis dynamometer, and an engine analyzer. The engine analyzer isn't an absolute necessity, but most shops have them because they're a great time-saver when it comes to tuning and troubleshooting. A new engine analyzer can cost from $5000 to $7000 or more. A used or rebuilt unit might be purchased for $1500 to $6000. Recognized leaders in the analyzer equipment field are Sun, Marquette, Hamilton Test, Kal Equip, Allen and Clayton.

An exhaust analyzer is needed to make the carburetor adjustments on the propane system. Some engines analyzers come with built-in exhaust analyzers. This usually adds $1000 to $2000 to the price of the engine analyzer. Separate exhaust analyzers can also be purchased for as little as a few hundred dollars, for a relatively simple unit, or up to several thousand dollars for a more sophisticated unit.

A chassis dynamometer is a unit consisting of two rollers placed in the floor and hooked to an electronic control module. The vehicle's drive wheels are placed over the rollers so that the engine can be run up to road speed in the shop. This allows the mechanic to tune the fuel system and ignition for peak performance under simulated driving conditions. A heavy-duty dyno costs from $10,000 up, plus installation.

An alternative to buying a lot of expensive equipment is to work out a deal with a local repair facility that already has the equipment you need. You do the conversions in your shop and then take the vehicles to Joe's Garage to finish the tuning. You really don't need a lift either as a good hydraulic floor jack can handle all the lifting that's required for most conversions.

FINDING GOOD MECHANICS

If you're not an expert in propane conversions, you'll need to hire someone, get training for yourself, or get training for your mechanics through your local LP-Gas Association or one of the various equipment manufacturer schools such as those conducted by Impco. Training is essential because most distributors won't sell parts to anyone who isn't trained. In addition, many states require mechanics who install propane fuel systems to be licensed or certified.

If you're opening a conversion center from scratch, it might be wise to hire an experienced propane mechanic before you do anything else. He can help you pick out the tools, equipment, and inventory you'll need. One way to obtain such skilled help is to hire mechanics away from an existing conversion center or fleet.

A typical conversion center might want to have four full-time mechanics on staff. Two would work on carburetion while the other two concentrate on tank installations. Even for smaller shops, it's a good idea to have two workers on duty because propane fuel tanks are quite heavy and they usually require two people to lift them into place.

Most shops usually pay the person who does the carburetor work a higher wage than the person who does tank installations because the carburetion requires more skill and less brawn. Most shops also pay their employees on an incentive plan to reduce the amount of time spent on a conversion.

INVENTORY

The amount of inventory you'll want to have on hand depends on the number of conversions you expect to do in a given period of time and the parts availability through your distributors. If you plan to handle a lot of work, you'd be wise to have plenty of parts on hand to avoid getting caught in the situation of having work to do, but no parts to do it with. Some shops will invest from $10,000 to as much as $30,000 or more in their basic equipment inventory. For a beginner, an inventory of 1500 to 2000 items for about $10,000 should fulfill parts needs. You should also stock some common tune-up parts such as spark plugs and contact points to supplement your inventory. These items, however, are readily available through any local auto parts store.

One very important consideration to keep in mind at all times with regard to parts is *buy quality equipment*. Propane fuel system components must meet NFPA safety standards, tanks must be ASME-certified and electrical components must be UL-approved for propane gas. Installing substandard or unapproved parts is illegal in most areas.

The National LP-Gas Association has a voluntary propane conversion certification program in which every conversion center should participate. The program certifies the people who do the conversions as well as the parts used.

LABOR AND PRICING

The average shop charges $800 to $1500 for a conversion that takes one or two mechanics about eight hours to complete. The carburetion parts can run from $400 to $500, and the tank can cost another $250 or so. This leaves a gross profit margin of around 40 to 50 percent.

These price figures apply to a normal market where gas prices are stable and service station lines are nonexistent. Should another gas crunch occur or gas prices suddenly skyrocket, conversion centers could increase their prices—and profits—dramatically.

PITFALLS

One tricky area to watch for is the insurance required for your business. Propane, although safer than gasoline, is still a dangerous and explosive fuel. Liability coverage is an absolute necessity.

Interestingly, insurance for the center should not be high. A standard garage liability coverage, commonly used by most repair shops, is sufficient. Insurance protection at reasonable rates is no problem for your customers either. Insurance on their vehicles should not change as a result of converting to propane.

Liability insurance is required on vehicles once they leave your center, and it must be paid for by you during the length of your warranty period. This coverage has a high premium; it is in the $1000 to $1500 per month range ($12,000 to $18,000 a year). This protection is required by insurance companies concerned about the quality of the conversions. If a conversion you did causes an accident, you could be held liable because of faulty workmanship. That's where your insurance protection comes in. You should also consider incorporating to protect your personal assets against any judgments that might result from your business. Incorporation adds to the start-up costs, but it is a wise precaution for any business.

Another area to thoroughly investigate before going into the conversion business is the law. Thirty states have strict licensing requirements for propane conversion centers. In California, Texas, and Florida, the mechanic who does the conversions must be licensed. Be sure you understand all the rules and regulations that apply to the conversion business in your area.

The greatest pitfall to avoid is that of sloppy installation work. Not only is it bad for business, but it can lead to lawsuits if negligence can be proven. Mechanics should be thoroughly trained and the work double-checked and guaranteed. Because the propane business has become such a hot investment in recent years, many fly-by-night artists are cashing in on the profits that are there for the taking. Eventually they're caught up with, but not until they've done damage. The industry has had an excellent safety record to date. Make sure you do your best to keep it that way.

One of the best ways to get started in the business is to get the training and know-how through a franchise program. The Dancey Company, P.O. Box 16188, 5402 E. Hanna, Tampa, Florida, has a one-week school costing $329 for fleet owners, mechanics, and would-be entrepreneurs. The course leads into a complete dealership program.

Dancey offers a complete turnkey franchise priced in the $100,000 range. Dancey's program is based on a retail conversion outlet with five employees. The franchise fee is $12,000 and an ongoing royalty of 6½ percent is charged on gross sales. The equipment included is valued at $14,000 and the inventory is valued at $40,000. The balance represents the additional working capital they estimate an operator needs to run the business.

Whether you want to invest $100,000 in Dancey's franchise or $2000 in NAMOCO's basic dealer package, getting into the propane conversion business can be a profitable and rewarding venture.

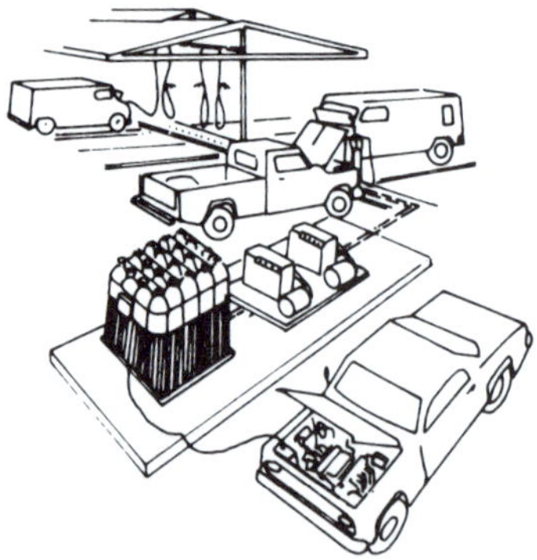

Chapter 12

Natural Gas

Although this book is primarily about propane, there is another gaseous fuel that is becoming increasingly popular as a low cost alternative to gasoline. That fuel is natural gas.

Most people are familiar with natural gas because many use it to heat their homes, heat water, and to cook meals. The use of natural gas for heating, cooking and industrial processes is so widespread that it accounts for approximately one-fourth of the nation's total energy needs.

In 1981, 165,000 natural gas wells in the United States produced nearly 20 trillion cubic feet of the gas. Industry experts agree that the United States should have adequate supplies of natural gas for at least the next decade and probably until the turn of the century. Beyond that it's anybody's guess. On the world scene, there is an over-abundance of natural gas. Canada, Mexico, and most of the oil-producing nations have more gas than they know what to do with. Until recently, much of it was simply flared off at the wellhead because the means of collecting and transporting it were nonexistent. Saudi Arabia, Indonesia, and others are now salvaging much of this gas as part of their oil-drilling operations for export to industrialized nations. The long term supply outlook for natural gas, therefore, seems to be favorable.

METHANE

Natural gas is composed primarily of methane (CH_4), with various amounts of propane, butane and other gases mixed in as it comes from the ground. The propane, butane, and other heavy gases are separated so that the remaining gas that enters the pipeline distribution network is 90 percent or more methane. Because methane is a colorless and odorless gas, a scent is added so that leaks can be smelled. This is the characteristic gas odor you smell when a pilot light blows out.

Methane is the same stuff that is often called *sewer gas* or *digester gas*, and is frequently a byproduct of municipal waste disposal systems. Methane is produced when anerobic bacteria decompose organic wastes.

Do-it-yourselfers can make their own methane fuel by filling an old drum with a mixture of animal manure and moist plant clippings. The bacteria in the manure will decompose the clippings and release methane. The gas is vented from the drum into any suitable storage tank, or it is routed to a stove or burner and burned. A tractor tire inner tube makes a good methane collector when connected to a methane generator drum with a length of tubing.

Methane can also be manufactured on a commercial basis from coal. This fact, combined with the proven geologic reserves of natural gas and the availability of bacteria-produced methane, gives natural gas a brighter supply outlook than any other gaseous fuel with the exception of hydrogen. Hydrogen is *the* most abundant element on Earth, and it can be extracted from water molecules by a process called *electrolysis*. Yet hydrogen remains one of those "someday if" fuels because it is still too expensive to produce to be an economical alternative for most applications. Until someone comes up with a way of making cheap hydrogen fuel, methane, and propane will be the short-term alternative to gasoline.

ADVANTAGES OF NATURAL GAS

The number one advantage of natural gas over conventional fuels is low price. See Table 12-1. In 1981, 100 cubic feet of natural gas (called a *therm*,) which contains about the same amount of energy as a gallon of gasoline, cost 40 cents to 45 cents. That's even less expensive than propane. Natural gas is currently the least expensive fuel available for almost every purpose. For those who have access to the sewer gas from municipal waste treatment facilities, the fuel is practically free.

In addition to its overwhelming economic advantage, methane has the same desirable performance characteristics of propane. It is a clean-burning fuel that produces a low amount of exhaust emissions and extends oil life, spark plug life, and engine life. It has an

Table 12-1. Gaseous Fuels.

	CNG	LNG	Propane
Chemical formula	CH_4	CH_4	C_3H_8
State when stored in fuel tank	gas 2400 psi	liquid 20 psi −260°F	liquid 200 psi
Octane rating (research)	130	130	112
Type of fuel tank	heavy steel cylinder	insulated steel tank	steel tank
Btu per gallon	—	83,000	91,000
Btu per cubic foot	1000	1000	2500
Btu per pound	21,500	21,500	19,750
Weight in lbs. per gallon		3.43	4.86
Ignition temp.	1170°F	1170°F	750°F
Price advantage compared to gasoline	2/3 less	2/3 less	1/3 to 1/2 less

octane rating even higher than that of propane (130 versus 112), and it burns cooler. This reduces exhaust valve wear that is sometimes a problem with propane-powered engines. Methane is also lighter than air. This means any fumes will drift upwards. In this respect, it is safer than gasoline or propane.

The big difference between methane and propane, however, is that methane remains a gas when pressurized; propane becomes a liquid. Although this can be considered an advantage in extremely cold weather (propane begins to lose its ability to vaporize as the temperature drops below −30 degrees), it greatly reduces the amount of fuel that can be squeezed into a fuel tank. That is methane's primary disadvantage.

COMPRESSED NATURAL GAS

Natural gas fuel systems for vehicles typically have very limited ranges because of the fuel storage problem. To get enough methane into a fuel tank to provide any range at all requires compressing the gas to very high pressures. Fuel used under such circumstances is called *compressed natural gas* or *CNG* for short.

CNG fuel tanks typically hold the gas at 2400 psi. That is about 10 times the pressure of a propane fuel tank. This creates several problems. First, the fuel tank must be very strong to withstand such high pressures. The tank will be heavy and bulky and that limits the amount of fuel a vehicle can carry. A typical CNG fuel tank is a 3½-gallon steel cylinder that weighs 140 pounds. Two such tanks could be mounted in the trunk of a passenger car to provide a driving range of 40 to 90 miles. That is a satisfactory range for many fleet vehicles, but it is not enough for a private car.

Because of the limited range, most CNG conversions are to dual-fuel systems so that the vehicle can still be driven on gasoline if necessary. The dual-fuel approach extends the useful range of the vehicle and takes the worry out of refueling should the CNG tank run low. But dual-fuel conversions sacrifice both performance and fuel economy because the engine must be tuned to run on both gasoline and methane. See Figs. 12-1 through 12-3.

There is yet another hitch to using compressed natural gas. Compressing the gas to 2400 psi requires a special high-pressure compressing station. These units can cost as much as $70,000. The high cost of compressing the fuel prices CNG conversions out of the reach of many small fleets, but it is still within the bounds of larger fleets or those with access to inexpensive fuel supplies. Most CNG conversions to date have been by municipal or gas company fleets.

Refueling CNG tanks can be done in two to five minutes if the tank is filled from a reservoir of high-pressure storage tanks (called a *cascade*). The vehicle could be refueled by pumping low-pressure

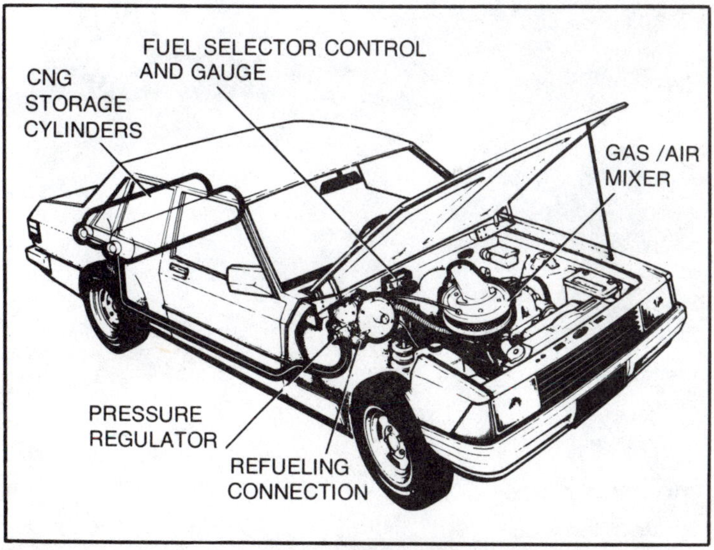

Fig. 12-1. The basic components of a CNG conversion.

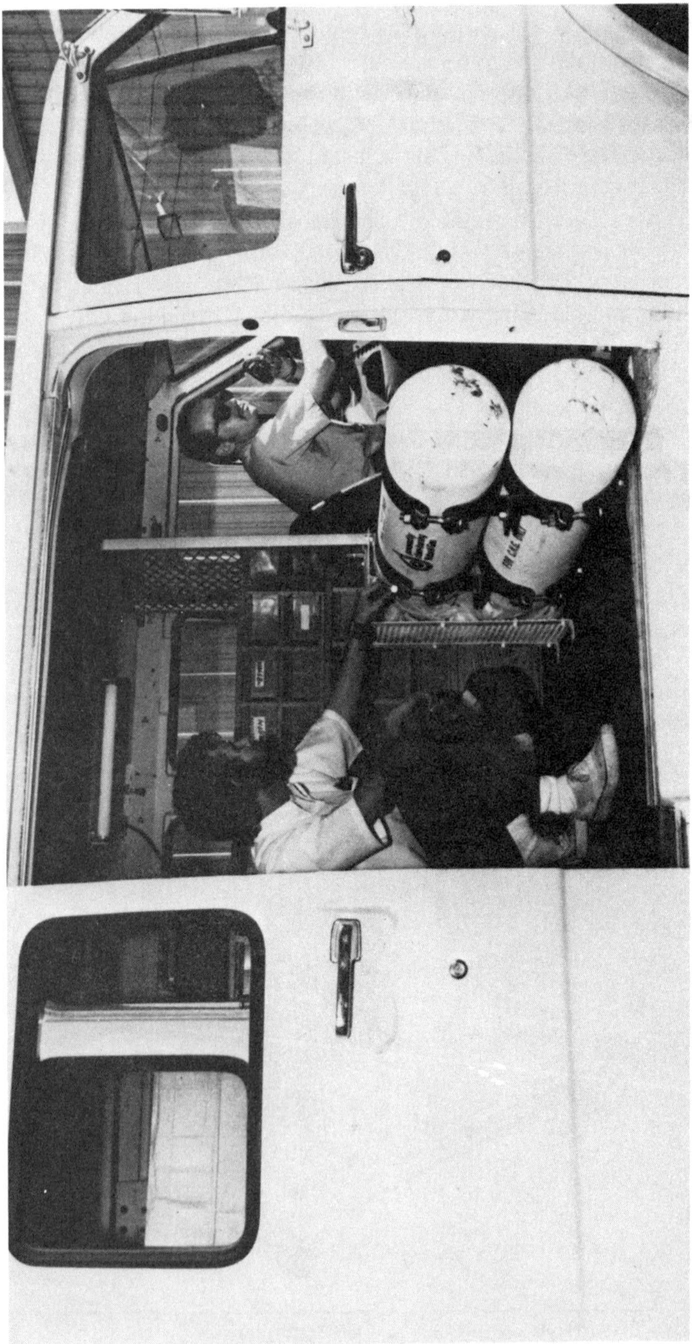

Fig. 12-2. A CNG-powered delivery van.

Fig. 12-3. A CNG-powered forklift truck at a Coors brewery.

natural gas direct from the gas line into the tank. This approach can take from 10 to 14 hours to reach the 2400 psi capacity of the fuel tank. This method used only where vehicles can be parked for overnight refueling (Figs. 12-4 and 12-5).

LIQUIFIED NATURAL GAS

An alternative to compressing natural gas to get more fuel into a given tank is to liquify the gas. If methane is cooled to −260 degrees, it will become a liquid. The liquified natural gas is called *LNG*.

When liquified, methane occupies about one-six-hundredth the volume it did when it was a gas. This allows more fuel to be stored inside a tank, compared to CNG, and eliminates the high pressures and compressing requirements. The catch is that special refrigeration equipment is needed to liquify natural gas. Such equipment can cost as much as $25,000, and in 1981 there were only about 105 LNG refueling facilities in the United States. None of which were available to the general public.

Liquifying natural gas is not as easy as it sounds. Before the gas can be refrigerated, all impurities must be removed. Moisture,

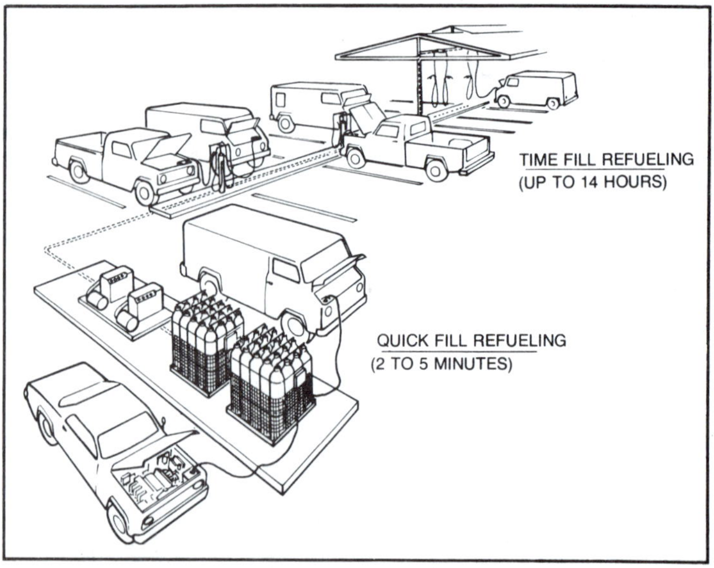

Fig. 12-4. CNG-powered vehicles can be refueled by quick filling, which is accomplished in 2 to 5 minutes by using a cascade of pressurized tanks, or time filling, which can take up to 14 hours using a central compressor to fill the tanks directly.

Fig. 12-5. A quick-fill cascade saves time by providing a reservoir of high pressure gas. The compressor fills the tanks. The fleet's vehicles are refueled from the cascade tanks.

dust, carbon dioxide and residual oils mixed in with the natural gas will turn to solid ice at low temperatures if they are not filtered out. Such impurities could clog fuel lines if not removed.

Once liquified, the super-cold LNG must be insulated to keep it from vaporizing. The fuel tank resembles a giant thermos bottle. Even the best tanks can hold fuel only six or seven days maximum, and then the fuel will start to warm up and evaporate. When that happens, the fuel must be vented to the outside to keep the tank from exploding. There is a time limit on how long a LNG fuel tank can sit before it starts to vent fuel. See Figs. 12-5 through 12-12.

As of late 1981, there was only one company making a LNG conversion kit for vehicles. Beech Aircraft of Wichita, Kansas, markets a LNG fuel system based on liquid fuel storage tanks it pioneered for the space program. The Beech LNG fuel tank has a stainless steel inner shell surrounded by aluminized mylar insulation and enclosed inside a permanent high vacuum. The tank is built to ASME codes and contains fluid valves and a fuel gauge. Normal tank pressure is 20 psi. When pressure builds up to 60 psi, fuel is vented to the outside through a pressure-relief valve. Fuel loss rate after seven days is said to be less than 1 percent a day. The tank holds 18 gallons of fuel (a 38-gallon tank is also available), and gives a small-sized passenger car a driving range of about 250 miles. Refueling takes about three minutes if the tank is still cold, or about seven minutes if it's empty.

Fig. 12-6. A Southern California Gas Company CNG-powered customer service truck.

Fig. 12-7. The refueling connection on this CNG-powered truck, converted by Dual Fuel Systems, Inc. is under the hood.

The actual cost of a LNG conversion depends on the size of the fleet being converted. Generally, it will be more expensive than a propane conversion because of the added expense of the insulated fuel tank. A ball-park estimate would be $2100 per vehicle plus the cost of the refrigeration equipment.

NATURAL GAS CONVERSIONS

In a CNG conversion, the fuel travels from the tanks to a high-pressure regulator that reduces the fuel pressure from 2400 psi to 60 psi. The fuel then passes to a low pressure regulator and is reduced to less than 1 psi before it goes to the carburetor or mixer. No converter is required because the fuel is already a gas. The pressure regulator incorporates a fuel shut-off to stop the flow of fuel when the engine is not running. The carburetor or fuel mixer is exactly the same as a propane unit except that it is recalibrated to supply a richer fuel mixture.

The installation of a CNG fuel system is basically similar to that of a propane conversion. For straight conversions, the gasoline system is replaced by the CNG system. For dual-fuel conversions, the CNG system is added to the gasoline system.

When CNG fuel tank cylinders are mounted in the trunk of a passenger car, the same safety rules apply as with propane. A vapor

Fig. 12-8. Part of Southern California Gas Company's CNG-powered fleet.

Fig. 12-9. An LNG-powered Ford Granada incorporating Beech Aircraft's LNG fuel system.

barrier must be provided to keep fumes out of the passenger compartment and a remote filler location must be used. Safety release valves must be provided and fuel lines must be rated at 2400 psi.

A CNG-powered car will typically have two or three trunk-mounted, fuel-tank cylinders; a pickup truck or van could have two to four and a larger truck could have as many as eight.

Engine modifications would be similar to those made for a propane conversion (more ignition advance, elimination of intake air and manifold preheating, lower thermostat, narrower spark plug gap, etc). The extent of the engine modifications would depend entirely on the application. For a straight CNG conversion, all the modifications just mentioned would be used to maximize fuel economy and performance. On a dual-fuel CNG conversion, the degree of modifications would depend on how frequently gasoline would be used as the main fuel. In most cases, the only change would be to slightly advance the ignition timing.

The only drawback to CNG dual-fuel conversions—like that of propane dual fuel conversions—is that performance and fuel economy are compromised because the engine must be tuned so that it can still operate on gasoline. The problem is that methane requires more spark advance than gasoline. Tuning the engine

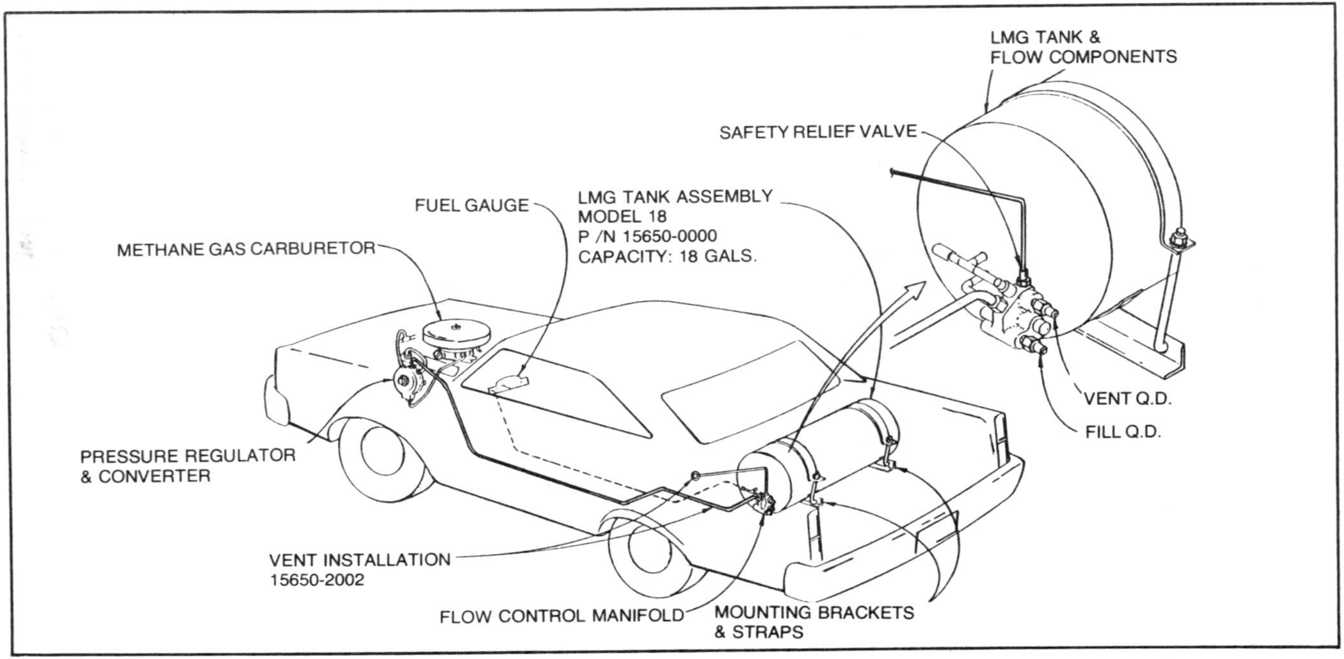

Fig. 12-10. Beech Aircraft's Liquid Natural Gas (LNG) system for automotive use. The super-cold fuel is stored in a huge vacuum bottle in the vehicle's trunk.

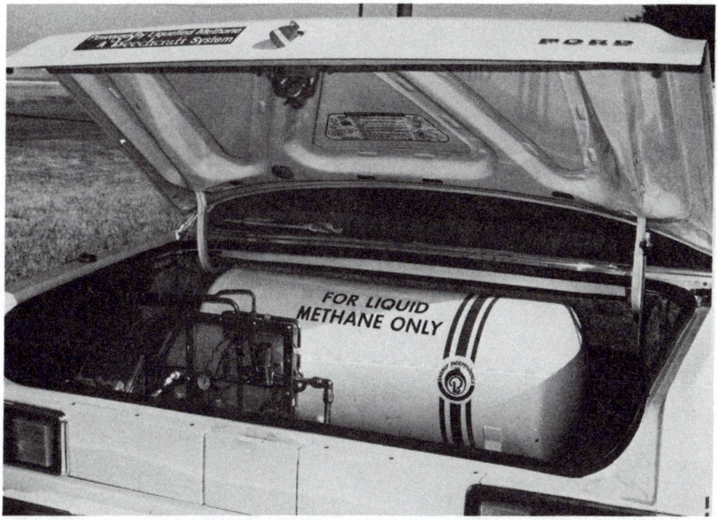

Fig. 12-11. In an LNG-powered vehicle, the natural gas or methane is kept at −26° inside an insulated vacuum tank. The tank can keep the fuel at this temperature for up to a week or more.

timing to methane can cause detonation problems when gasoline is used.

In the past, the usual approach was to bump the timing a couple of degrees and accept less than optimum performance when running on methane. Dual Fuel Systems Inc. now has an electronic "black box" it calls Dual Curve Ignition that will automatically change the ignition curve when the engine is switched from gasoline to methane. The box provides the optimum timing for both fuels for increased performance and fuel economy. An easy change over of wires at the ignition coil is all that's needed to install the electronic timing control.

As for LNG conversions, the basics are the same as for CNG conversions. The only differences are:

- a single insulated fuel tank is used.
- the fuel system working pressure is 20 psi.
- a converter is necessary to change the super-cold LNG into a gas before it reaches the carburetor or mixer.

FLEET EXPERIENCE WITH CNG

Dual Fuel Systems Inc., a Culver City, California-based subsidiary of the Pacific Lighting Corp., has converted thousands of vehicles to CNG dual-fuel use. Dual Fuel helped the Los Angeles

Fig. 12-12. Under the hood, an LNG conversion looks much like a propane conversion. The fuel mixer is mounted on an offset adaptor to provide needed hood clearance. The converter is at the left side of the engine compartment next to the firewall.

County Sanitation District convert its fleet so the District could make use of its digester gas.

In 1975, Dual Fuel converted 120 vehicles for Home Savings and Loan Association, a Los Angeles financial institution. At that time, gasoline was selling for only 45 cents a gallon. Home recovered its initial conversion costs in less than two years and has saved thousands of dollars in fuel bills every year since.

Many gas utility companies across the country have switched to CNG or they are in the process of changing over. The Southern California Gas Company has most of its 4700 customer service vehicles converted to CNG dual fuel. Some of the others include Pacific Gas and Electric of California, Iroquois Gas Company of New York, Southwest Gas Company of Nevada, Leclede Gas Company, and Cities Service Gas Company of Missouri, and Piedmont Gas Company of North Carolina.

Many commercial fleets are also converting to CNG to take advantage of the tremendous fuel savings that are possible with natural gas. The Yellow Cab companies of Ventura, Santa Barbara, Anaheim, San Luis Obispo and Redwood City, California have converted to CNG, as have fleets for Disneyland, Data Transportation, Johns-Manville, Kaiser Steel, Northrup Aircraft, IBM, and Canteen Corporation.

Vehicles in the municipal fleets of Coalinga and Long Beach, California, and Lawrenceville and Monroe, Georgia have also been converted to CNG. And the city of San Luis Obispo, California is operating its minibus transit system with dual-fuel power.

Pacific Northwest Bell of Seattle, Washington conducted a test involving 67 CNG-powered vans with dual-fuel capability to test the success of the conversions. The results were "very positive" with the fuel savings alone paying for the conversion within 30 months.

CNG cost Bell 43 cents per therm (equivalent to a gallon of gasoline) at the time of the test. Gasoline was selling for $1.25 a gallon. Fitted with twin 80-pound CNG fuel tank cylinders, the converted vans carried fuel equal to about 6½ gallons of gasoline. This provided a driving range of about 90 miles on CNG. Bell installed both a quick-fill station capable of refueling a van in two to five minutes, and a time-fill station that could refuel all 67 trucks in 10 to 14 hours overnight. The quick-fill station cost about $28,000 and the time-fill station cost about $67,000. Experts say a fleet owner should estimate about $1000 to $2000 per vehicle for a time-fill compressing station.

Officials of the Eaton School District, located 7 miles north of Greeley, Colorado, decided to switch to CNG rather than propane for two reasons: CNG is cheaper and it's safer. The District purchased CNG conversion equipment for 14 buses—including one compressor station for refueling—for a price of $52,396. Additional costs included freight, a concrete pad for the compressor, gas detectors, electrical and gas hook-ups, and contracting a private mechanic to install the equipment. All this brought the total conversion cost to $68,000. Officials estimate the District will recover the costs within 2½ years.

The Eaton School District reports that fuel consumption and performance are about the same on CNG as gasoline. The districts buses use four different sized engines: Ford 330- and 361-cubic-inch engines, and Chevrolet 350- and 366-engines. All have performed well on CNG, according to officials, and the drivers report experiencing no problems.

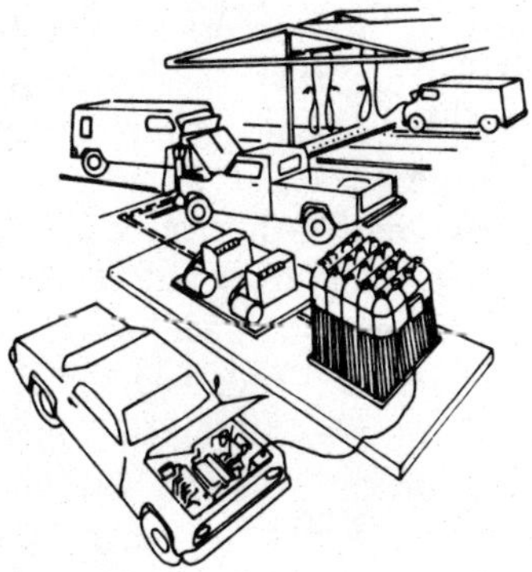

Glossary

air valve carburetor—A type of carburetor commonly used in straight propane conversions. A spring-loaded diaphragm is used to sense air flow through the carburetor. The diaphragm is connected to a gas metering valve that allows propane to be drawn into the carburetor. As air flow increases, the diaphragm rises and pulls open the gas valve in proportion to the amount of propane needed.

balance line—A vent that runs between the carburetor air horn and the atmospheric vent on the converter-regulator. A balance line is usually required on turbocharged engines or those with high air intake restrictions in order to compensate for the pressure differences that would otherwise upset the fuel flow.

Btu—Abbreviation for British thermal unit. A Btu is a measure of heat energy. One Btu equals the amount of heat required to raise the temperature of 1 pound of water 1 degree Fahrenheit. The energy content of motor fuels is often expressed in Btu per gallon, per pound, or per cubic foot. Propane, for example, contains around 2500 Btu per cubic foot.

butane—A gaseous hydrocarbon similar to propane. It has the chemical formula C_4H_{10}. Butane is primarily a by-product of crude oil refining. It can also be found in natural gas. Butane has a lower octane rating than propane, and it is not used as a motor fuel because of its limited availability compared to propane.

Small amounts of butane might be present in LP-gas fuels, but its primary use is in the chemical industry.

carbon monoxide—A harmful air pollutant found in vehicle exhaust that results from the incomplete combustion of hydrocarbon-based motor fuels. Carbon monoxide (CO) can be deadly because it combines with the blood more readily than oxygen. The formation of carbon monoxide in the exhaust can be minimized by avoiding over-rich air/fuel mixtures in the carburetor, and by adding a catalytic converter to the exhaust pipe.

carburetor—Any device that is used to mix air and fuel as it is drawn into an engine. The metering within the carburetor determines the air/fuel ratio of the mixture. A throttle controls engine speed by restricting the amount of air that can be drawn into the engine. Propane enters the carburetor as a "dry" gas whereas gasoline enters as a liquid.

catalytic converter—An air pollution control device that reduces carbon monoxide and hydrocarbon exhaust emissions by oxidizing the pollutants. Extra oxygen is added to the exhaust gases by an air pump ahead of the converter. A catalyst such as platinum or palladium causes the hot gases and oxygen to react. The result is that carbon monoxide and unburned hydrocarbons are converted to harmless carbon dioxide and water. Catalytic converters are standard equipment on all automobiles built since 1975. The catalyst can be ruined by the use of leaded gasoline or by excessively rich air/fuel mixtures.

CNG—Abbreviation for compressed natural gas. CNG is largely methane (CH_4) and can be used as a gaseous motor fuel like propane.CNG has a higher octane rating than propane, and it is stored as a highly compressed gas at about 2400 psi. It is a popular alternative fuel due to its availability and low price.

compression ratio—A term that describes the volume of an engine's combustion chamber at the end of the compression stroke compared to the volume at the beginning. It is the amount by which the air/fuel mixture is compressed prior to combustion. Most of today's engines have compression ratios around 8 to 1. Higher compression ratios increase fuel economy and performance, but require higher octane fuel to resist detonation.

converter-regulator—Also called the vaporizer-regulator or converter, it is the device in a propane fuel system that converts high-pressure liquid propane to a low-pressure gas before it enters the carburetor. The unit receives high-pressure liquid fuel from the fuelock/filter and reduces the pressure in two stages. When the propane is allowed to expand from 150-psi-tank pressure to atmospheric pressure, it changes from a liquid to a gas. This causes a drop in temperature. To compensate for the cooling and to aid vaporization, hot water from the engine's cooling system is circulated through the converter to heat the gas. The regulator then allows the gas to enter the hose that runs to the carburetor when a partial vacuum is created by the engine cranking or running. The regulator shuts off the fuel flow when the engine is not running.

detonation—A condition where the fuel ignites spontaneously inside the engine's combustion chamber. Instead of burning in a controlled fashion, the mixture explodes erratically and produces an audible knock or ping (sometimes called ***spark knock***) that is most noticeable on acceleration under load. Light detonation will reduce fuel mileage and it can sometimes damage head gaskets. Heavy detonation will damage rings, melt pistons, and often ruin the entire engine. Detonation is usually caused by over-advanced ignition timing, low-octane fuel, excessive compression (due to carbon buildup), or overheating of the engine or incoming air.

dual-fuel—Any fuel system that operates on a primary fuel and can be switched over to a backup fuel as needed. In automotive applications, the dual-fuels are typically propane-gasoline or CNG-gasoline. In stationary engines, natural gas-propane or synthetic gas-propane combinations are often used. The basic idea is to operate the engine on the less expensive fuel as much as possible, and use the backup fuel only when the primary fuel is unavailable.

EGR—Abbreviation for exhaust gas recirculation. EGR is an emission control technique that reduces the amount of nitrous oxides (NOX) in a vehicle's exhaust. An EGR valve allows a small amount of exhaust gas to be drawn into the intake manifold where it dilutes the incoming air/fuel mixture. This reduces combustion temperatures to below 2500° F to greatly reduce the formation of NOX. The EGR system has been standard on most automobiles since 1973.

fuelock/filter—A combination fuel shutoff valve and filter located in the high pressure fuel line between the propane fuel tank and converter-regulator. On vacuum models, the vacuum created by the engine cranking or running causes the fuelock to open so that fuel can flow to the converter-regulator. An electrical solenoid blocks the fuel flow when the ignition is switched off. The filter traps any sediment in the fuel or rust flakes from the tank before they can reach the converter-regulator.

gasoline—A mixture of various liquid hydrocarbons derived mainly from crude oil.

governor—Any device that limits engine speed to a predetermined maximum. Limiting engine rpm reduces stress, conserves fuel and prolongs engine life. Governors are typically used on trucks and fleet vehicles.

hydrocarbons—Gasoline, propane, and methane are all hydrocarbon fuels; their molecules contain hydrogen and carbon atoms. When burned completely, any hydrocarbon fuel will produce carbon dioxide and water vapor—both harmless substances. Unfortunately, achieving complete combustion in an internal combustion engine is next to impossible. The result is unburned hydrocarbons (soot) in the exhaust. Unburned hydrocarbons (HC) are considered to be air pollutants. Excessive HC emissions can result from ignition misfiring, oil burning, or over-rich air/fuel mixtures. HC exhaust emissions can be greatly reduced by a catalytic converter. Propane and methane typically produce very low HC emissions.

ignition curve—This is the term that describes the rate of change of the ignition timing with respect to engine rpm and load. As engine speed increases, the ignition timing must advance to allow sufficient time for the fuel to burn. Over-advanced or over-retarded timing will reduce mileage and performance. Centrifugal advance is added by small flyweights inside the distributor. Vacuum advance is added by a diaphragm attached to the distributor housing. Increasing engine speed adds centrifugal advance and increasing engine intake vacuum adds vacuum advance. The two combine to form total advance and to determine the advance curve. In engines with computerized ignition systems, the computer determines spark timing according to the input from its various engine sensors. The timing is advanced a predetermined amount according to the set of conditions existing at any given instant. The increased timing control a computer offers improves fuel economy and reduces exhaust emissions.

LP-gas—Also called LPG, it is an abbreviation for liquified petroleum gas. Although it is often used interchangeably with propane, there are actually four grades of LP-gases: *commercial propane* is primarily propane and propylene; *commerical butane* is primarily butane and butylene; *commercial B-P* mixtures are combinations of butane, butylene, propane and propylene; and *HD-5 propane* is supposed to contain at least 90 percent propane by volume and not more than 5 percent propylene. HD-5 propane is the only grade of LP-gas suitable for use as a motor fuel.

methane—A high-octane gaseous fuel, methane (chemical formula CH_4) is the major component of natural gas. Unlike propane or gasoline, methane can be produced from renewable resources such as biomass or from urban wastes. The anerobic digestion of organic matter by bacteria releases methane (also called biogas or sewer gas). This can be collected, compressed for storage, and then used as a motor fuel.

mixer—A term used to describe a gaseous carburetor, typically the gaseous carburetor that sits atop the regular gasoline carburetor in a dual-fuel conversion. The mixer mixes propane or natural gas with air.

natural gas—The name given to the gas recovered from wells in the ground. Natural gas is a mixture of various gases. It is primarily methane with varying amounts of ethane, butane, propane, and other hydrocarbons. Like propane, natural gas is a clean-burning, efficient motor fuel. It can be stored under pressure or liquified (*LNG* is liquified natural gas). Compressed natural gas (*CNG*) is typically used in automotive applications rather than LNG because of reduced handling problems.

nitrous oxides—A harmful exhaust pollutant that is formed when high combustion temperatures combine nitrogen and oxygen from the air to create various nitrogen-oxygen (NOX) compounds. NOX formation in an engine is controlled by the EGR system. It is reduced in the exhaust by a reduction catalyst (3-way catalytic converters in 1981 and newer automobiles).

octane—A term used to describe a fuel's resistance to detonation. The higher the numerical octane rating the more resistant the fuel is to detonation. Octane is rated according to several different scales: research (the highest), motor (the lowest), and pump (the average of research and motor ratings). Propane, for example, has a research octane rating of 112, methane is 120 and gasoline is 91 to 100. A higher octane allows the use of higher engine compression for improved performance and fuel economy.

propane—A high-octane, clean-burning gaseous fuel, propane has a chemical formula of C_3H_8. It is a by-product of crude oil refining. It is also found in natural gas. Sometimes called LP-gas, propane is stored as a high-pressure liquid (150 to 175 psi). It's main attraction as a motor fuel is its price advantage over gasoline. Other benefits include extended oil life, extended spark plug life, and longer engine life, and a less precarious supply situation.

turbocharger—A mechanical device for boosting intake manifold pressure to increase engine horsepower. A turbocharger is used mainly on trucks, performance cars, or for added high-altitude power. The turbo is positioned between the carburetor and intake manifold. The engine's exhaust is routed through a small turbine. This turns a second turbine that forces more air/fuel mixture into the engine as rpm increase.

valve recession—A condition where the engine's valve seats wear rapidly. It is caused by a lack of "lubrication" from combustion by-products, and is sometimes a problem in older vehicles that are operated on unleaded gasoline, propane, or methane. It is not a problem in newer engines that are designed to run on unleaded gasoline because these engines have hardened valve seats. In older engines, valve recession can be prevented by installing hardened valve seats or valves.

Sources of Information

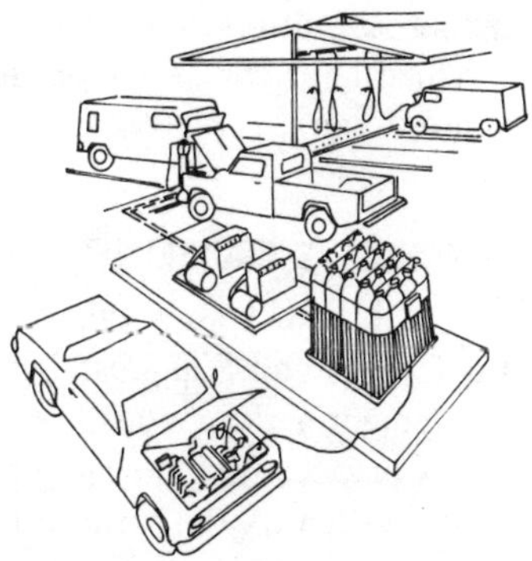

This section lists the major sources of information and suppliers of propane fuel, carburetion equipment, fuel tanks, fuel handling equipment, and storage tanks. There are also sources for propane conversions, training, and service.

OBTAINING PROPANE FUEL

Check in your local telephone book listings under Propane; LP-Gas; Liquified Petroleum Gas; Gas: Liquified Petroleum; Gas: Bottled & Bulk. There are about 25,000 propane dealers nationwide.

Contact your state or regional LP-Gas Association for a list of propane dealers near you.

If you're interested in making large bulk purchases of fuel for a fleet application, check with the major propane marketers listed in this section. You might find it most economical to become a dealer yourself.

VEHICLE CONVERSIONS AND SERVICE

Check in your local telephone book listings under Propane Equipment and Supplies; LP-Gas Equipment and Supplies; Gas; Equipment and Supplies; Propane Conversion and Service. There are about 3000 conversion centers nationwide.

The National LP-Gas Association publishes a directory that lists hundreds of conversion centers around the country. This directory is available free of charge on request. Requests must be made on your business letterhead.

Check with your state or regional LP-Gas Association for a list of conversion centers near you.

The major propane marketers listed in this section maintain numerous outlets regionally or nationally. These centers can convert cars, trucks, RV's or fleets. Write the propane marketers nearest you.

PURCHASING EQUIPMENT

Check in your local telephone book listings under Propane Equipment and Supplies; LP-Gas Equipment and Supplies; Gas: Equipment and Supplies.

Contact any of the equipment distributors listed in this section.

You can try the carburetion/equipment and fuel tank manufacturers listed in this section, but few manufacturers will sell direct to the public. They prefer to deal with distributors, conversion centers, or fleets with trained personnel.

Some conversion centers also sell parts and equipment. For a conversion center near you, try your local telephone book listings, contact your state or regional LP-Gas Association, or write to the National LP-Gas Association for their directory.

TRAINING

Contact your state or regional LP-Gas Association. Most associations conduct periodic schools and seminars for mechanics who want to learn how to do propane conversions.

Many conversion centers also offer training (usually for a fee) to fleet mechanics or persons interested in becoming distributors or installers. Contact the major propane marketers listed in this section or try a conversion center near you.

Many carburetion equipment manufacturers, such as Impco, conduct schools and seminars. Contact the manufacturers for times and locations.

SAFETY INFORMATION

Handling and storing propane fuel as well as installing a propane fuel system comes under federal and state fire regulations. The standards for these regulations are described in a booklet (NFPA #58) published by the National Fire Protection Association.

CURRENT MARKET INFORMATION

For fuel prices, fuel availability, or other information about propane or propane equipment, contact your state or regional LP-Gas Association, or any of the major propane marketers listed in this section.

Subscribe to one of the industry publications: *LP-Gas Magazine* or *Butane-Propane News*.

INFORMATION ON NATURAL GAS

Contact your local gas utility company.

Some of the carburetor equipment manufacturers also make equipment for compressed natural gas applications.

Contact Dual Fuel Systems, Inc., 6140 Bristol Parkway, Culver City, CA 90230; phone (213) 649-1062 or (800) 421-1171. Dual Fuel is a subsidiary of Pacific Lighting Corp., and has converted thousands of fleet vehicles to CNG/gasoline.

American Gas Association, 1515 Wilson Blvd., Arlington, VA 22209; phone (202) 841-8400.

Beech Aircraft Corp., Alternative Energy Div., P.O. Box 85, Wichita, KS 67201. Beech has developed an LNG system for car and truck fleets.

LP-GAS ASSOCIATIONS

National LP-Gas Association
1301 West 22nd St.
Oak Brook, IL 60521
(312) 986-4800

Alabama LP-Gas Assn.
660 Adams Ave.
Montgomery, AL 36104
(205) 264-9630

Arizona LP-Gas Assn.
(c/o NLPGA regional office)
P.O. Box 160232
1507 21st St.
Suite 109
Sacramento, CA 95816
(916) 441-6102

Arkansas LP-Gas Assn.
421 Donaghey Bldg.
Little Rock, AR 72201
(501) 374-8396

Colorado LP-Gas Assn.
6767 E. 39th Ave.
Denver, CO 80207
(303) 321-8523

Florida LP-Gas Assn.
P.O. Box 897
540 S. Maitland Ave.
Maitland, FL 32751
(305) 644-2063

Georgia LP-Gas Assn.
3131 Maple Dr. NE
Suite 101
Atlanta, GA 30305
(404) 233-4178

Idaho Propane Gas Assn.
(c/o NLPGA regional office)
P.O. Box 160232
1507 21st St.
Suite 109
Sacramento, CA 95816
(916) 441-6102

Illinois LP-Gas Assn.
3085 Stevenson Drive
Springfield, IL 62703
(217) 529-3943

Indiana LP-Gas Assn.
1111 E. 54th St.
Suite 149
Indianapolis, IN 46220
(317) 251-5141

Iowa LP-Gas Assn.
(c/o NLPGA regional office)
7515 Wayzata Blvd.
Minneapolis, MN 55426
(612) 544-3339

Kansas LP-Gas Assn.
701 Jackson
Room 208
Topeka, KS 66603
(913) 354-1749

Kentucky LP-Gas Assn.
P.O. Box 592
109 Thistle Road
Frankfort, KY 40602
(502) 875-2686

Liquified Petroleum Gas Assn. of New England
715 Boylston St.
Boston, MA 02116
(617) 266-6800

Louisiana LP-Gas Assn.
P.O. Box 2754
1915 Stubbs Ave.
Suite 101
Monroe, LA 71201
(318) 322-6547

LP-Gas Assn. of Maryland
(c/o NLPGA regional office)
1800 N. Kent St.
Arlington, VA 22209
(703) 525-5133

Michigan State LP-Gas Assn.
(c/o NLPGA)
1301 W. 22nd St.
Oak Brook, IL 60521
(312) 986-4807

Minnesota LP-Gas Assn.
7515 Wayzata Blvd.
Rm. 101
Minneapolis, MN 55426
(612) 544-3339

Mississippi LP-Gas Assn.
170 E. Griffith St.
Jackson, MS 39202
(601) 354-4077

Missouri LP-Gas Assn.
521 E. High St.
Jefferson City, MO 65101
(314) 635-0248

Montana-Wyoming LP-Gas Assn.
(c/o NLPGA regional office)
6767 E. 39th Ave.
Denver, CO 80207
(303) 321-8523

Assn. of Nebraska LP-Gas Dealers
Terminal Bldg.
Suite 518
Lincoln, NE 68508
(402) 475-3996

Nevada Liquified Gas Dealers Assn.
2206 Birch St.
Carson City, NV 89701
(702) 882-6699

New Jersey LP-Gas Assn.
(c/o NLPGA regional office)
1800 N. Kent St.
Arlington, VA 22209
(703) 525-5133

New Mexico LP-Gas Assn.
(c/o NLPGA regional office)
6767 E. 39th Ave.
Denver, CO 80207
(303) 321-8523

New York State LP-Gas Assn.
(c/o NLPGA regional office)
1800 N. Kent St.
Arlington, VA 22209
(703) 525-5133

North Carolina LP-Gas Assn.
10 Buck Jones Rd.
Raleigh, NC 27606
(919) 851-1934

North Dakota LP-Gas Assn.
(c/o NLPGA regional office)
7515 Wayzata Blvd.
Minneapolis, MN 55426
(612) 544-3339

Ohio LP-Gas Assn.
Rt. 2
Warsaw, OH 43844
(614) 824-4114

Oklahoma LP-Gas Assn.
2910 N. Walnut
Suite 114A
Oklahoma City, OK 73105
(405) 525-9386

Oregon Liquified Gas Assn.
(c/o NLPGA regional office)
P.O. Box 160232
1507 21st St.
Suite 109
Sacramento, CA 95816
(916) 441-6102

Pennsylvania LP-Gas Assn.
(c/o NLPGA regional office)
1800 N. Kent St.
Arlington, VA 22209
(703) 525-5133

South Carolina LP-Gas Assn.
(c/o NLPGA regional office)
3301 Buckeye Road
Suite 104
Atlanta, GA 30341
(404) 451-2017

South Dakota LP-Gas Assn.
P.O. Box 1580
Rapid City, SD 57709
(605) 343-6917

Tennessee LP-Gas Assn.
(c/o NLPGA regional office)
3301 Buckeye Road
Suite 104
Atlanta, GA 30341
(404) 451-2017

Texas LP-Gas Assn.
P.O. Box 9925
Austin, TX 78766
(512) 836-8620

Utah Liquified Gas Assn.
(c/o NLPGA regional office)
6767 E. 39th Ave.
Denver, CO 80207
(303) 321-8523

Virginia LP-Gas Assn.
(c/o NLPGA regional office)
1800 N. Kent St.
Arlington, VA 22209
(703) 525-5133

Washington Propane Assn.
(c/o NLPGA regional office)
P.O. Box 160232
1507 21st St.
Suite 109
Sacramento, CA 95816
(916) 441-6102

West Virginia LP-Gas Assn.
(c/o NLPGA regional office)
1800 N. Kent St.
Arlington, VA 22209
(703) 525-5133

Western Liquid Gas Assn.
P.O. Box 160232
1507 21st St.
Suite 109
Sacramento, CA 95816
(916) 441-6102

Wisconsin LP-Gas Assn.
(c/o NLPGA regional office)
7515 Wayzata Blvd.
Minneapolis, MN 55426
(612) 544-3339

FOREIGN LP-GAS ASSOCIATIONS

Propane Gas Assn. of Canada
500 4th Ave SW
Suite 1202
Calgary, T2P 2V6
Alberta, Canada
(403) 263-0450

The Japan LP-Gas Assn.
Hiroshi Takiguchi
1-1-21 Nichishinbashi
Minato-Ku, Tokyo, Japan

Asociacion Nacional De Distribuidores De Gas L.P.A.C.
V. Morales Galindo
Filadelfia No. 119-100
Mexicao City 18, D.F.
Mexico

MAJOR PROPANE MARKETERS

AmeriGas: LP-Gas Division
P.O. Box 965
Valley Forge, PA 19482

Aminoil USA Inc.
P.O. Box 94193
Houston, TX 77019

Buckeye Gas Products Co.
320 S. Boston Bldg.
Tulsa, OK 74103

Cal Gas Corp.
P.O. Box 28397
Sacramento, CA 95828

Cities Service Co.
P.O. Box 300
Tulsa, OK 74102

Empire Gas Corp.
P.O. Box 303
Lebanon, TX 77002

Getty Refining & Marketing Co.
P.O. Box 1650
Tulsa, OK 74102

National Propane Co.
6917 Collins Ave.
Miami Beach, FL 33141

Northern Propane Gas Co.
2223 Dodge St.
Omaha, NE 68102

Pargas, Inc.
P.O. Box 67
Waldorf, MD 20601

Petrolane LP-Gas Division
P.O. Drawer 1410
Long Beach, CA 90801

Pyrofax Gas Corp.
P.O. Box 2521
Houston, TX 77001

Suburban Propane Gas Corp.
P.O. Box 206
Whippany, NJ 07981

Texgas, Union Texas Petroleum
P.O. Box 2120
Houston, TX 77001

Thermogas Inc.
P.O. Box 4055
Des Moines, IA 50333

Tropigas Int'l Corp.
P.O. Box 341218
Coral Gables, FL 33114

Vangas Inc.
P.O. Box 12104
Fresno, CA 93776

EQUIPMENT DISTRIBUTORS

Northeast

Ace Bottled Gas Co.
Monmouth Road
Wrightstown, NJ 08562

Ace Bottled Gas Co.
1111 Route 37 West
Toms River, NJ 08573

Checkeye LPG Carburetion Inc.
651 Pittsburgh St.
Springdale, PA 15144

Continental Propane Equipment
24th Wharton St.
Pittsburgh, PA 15203

Eastern Propane Corporation
255 Oak Ridge Road
Oak Ridge, NJ 07438

Energy Trends Inc.
180 Research Parkway
Meriden, CT 06450

Flame-Rite Gas Inc.
3927 East Newport Rd.
Gordonville, PA 17529

Fleet Propane Services Inc.
2525 River Rd.
Niagara Falls, NY 14304

Garden State Propane Gas Corp.
3129 Route 46
Parsippany, NJ 07054

Gas Service Inc.
602 Raleigh St.
Bluefield, WV 24701

Glover Bottled Gas Corp.
675 Medford Road
Patchogue, NY 11772

Irish Propane
1177 Sycamore St.
Buffalo, NY 14211

Kellam Propane Gas Co.
P.O. Box 246
Belle Haven, VA 23306
(804) 442-5811

Liberty Gas Inc.
P.O. Box 548
Waldorf, MD 20601

New Jersey Propane Corp.
4700 S. Clinton Ave.
South Plainfield, NJ 07080

New York Propane Corp.
175 Price Parkway
Farmingdale, NY 11735

North Jersey Propane Gas Corp.
163 Beaver Brook Rd Box 525
Lincoln Park, NJ 07035

Propane Corporation of America
175 Price Parkway
Farmingdale, NY 11735

Propane Equipment Corp.
11 Apple St. Box 836
Tinton Falls, NJ 07724

Propower Associates Inc.
25 W. Olive St.
Westville, NJ 08096

Ramco Sales & Service Inc.
Bldg. S-2 Ryan Ave. P. O. Box 407
Westville, NJ 08093
(609) 456-8220

Schagrin Gas Co.
1000 N. Broad St.
Middletown, DE 19709

Shawleys Superior L P Gas
Maugans Ave at I-81 Martin Rd.
Hagerstown, MD 21740

Star Gas Service
701 Hampton Road
Cherry Hill, NJ 08002
(609) 662-7500

Utility Propane Company
One Elizabethtown Plaza
Elizabeth, NJ 07207

Southeast

Auto & Truck Propane Co.
2027 N. Dixie Hwy.
Pompano Beach, FL 33060

Country Gas Service Inc.
P.O. Box 419
Sumiton, AL 35148

The Dancey Company Inc
5402 E. Hanna Avenue
Tampa, FL 33687

Dealers LP Equipment Company
3142 Connahbrook Drive
Memphis, TN 38116

Dexter LP Gas Co Inc.
3710 Western Bypass
Montgomery, AL 36108

Gardner-Marsh Gas Equipment Co.
Box 30699
Raleigh, NC 27622

Gibson Homegas Inc.
3542 Lamar Ave.
Memphis, TN 38116

Hamilton-Willoughby Propane
318 Wiregrass St-Box 208
Homerville, GA 31634

Hillsboro Bottled Gas Company
5307 E. Hanna Avenue
Tampa, FL 33610

Isometrics Inc.
Box 660
Reidsville, NC 27320

LPG Motor Fuels Inc.
3907 Lawrenceville Hwy
Tucker, GA 30084

Moulden Supply Co
3600 Hwy 80 West Box 1247
Jackson, MS 39205

Petro Systems International
402 Harding Industrial Dr.
Nashville, TN 37211

Propane Corporation of America
3126 Oakcliff Industrial St.
Doraville, GA 30040

Propane Motor Sales Co.
1111 Grecade St.
Greensboro, NC 27402

Propane Trucks Inc.
P.O. Box 340
Apex, NC 27502

Rebel Propane Gas Co. Inc.
8017 W. Beaver St.
Jacksonville, FL 32205

Rutherford Equipment Inc.
7173 G. Covington Hwy Box 549
Lithonia, GA 30058

Sawyer Gas of Jacksonville Inc.
7162 Phillips Hwy
Jacksonville, FL 32216

Squibb-Taylor Inc.
3119 Fleetbrook Dr.
Memphis, TN 38116

Suburban Mfg. Co.
Box 399 North Broadway
Dayton, TN 37321

Lowry Tims Co. Inc.
Hwy 61 South P.O. Box 249
Cleveland, MS 38732

White River Distributors
River Stadium Drive
Batesville, AR 72501

North Central

Acme Propane Carburetion Inc.
Hiway 169 No
Mankato, MN 56001

Altura Hardware
Altura, MN 55910

American Dual Fuel Inc
3909 W. Belt Hwy
Middleton, WI 53562

Sam Annis and Company
807 Rainbow Drive
Waterloo, IA 50704

Auto Energy Inc.
7404 Washington Ave. So
Eden Prairie, MN 55344

Burling Sales Associates
3030 Harbor Ln
Minneapolis, MN 55441

Crystal Cooperative
721 West Humphrey
Lake Crystal, MN 56055

Economy Gas Co.
Box 561
Litchfield, MN 55355

Gas Supply Inc.
2238 Edgewood Ave. S.
Minneapolis, MN 55426
(612) 545-1466

Hills Gas & Appliance Co.
414 E. Blvd N. Box 210
Rapid City, SD 57709

Hills Gas & Appliance Company
Riverview Industrial Park
Yankton, SD 57078

Lees Propane & RV Service
625 West 'O'
Lincoln, NE 68528

LPF Equipment Company Incorporated
517 East 5th Box 1281
Grand Island, NE 68802

Northern Propane Gas Co.
2223 Dodge
Omaha, NE 68102

Northwest LP Gas
219 S. Union
Mora, MN 55051

Propane Motor Corp
758 Schmidt Rd.
Davenport, IA 52802

Unigas Universal Energy Corp.
1301 E. Main Street
Waukesha, WI 53186

Versatec Distributors
1305 E. Main Street
Waukesha, WI 53186
(414) 542-4640

Waterloo Service Company
Box 300 1402 Logan Avenue
Waterloo, IA 50704

Mid-Continent

Almac Engineering Inc.
20769 Mound Road
Warren, MI 48091

Blocker Propane Equipment Co.
417 Madison Avenue
Indianapolis, IN 46225

Blocker Propane Equipment Co.
15 E. South Street
Indianapolis, IN 46225

East Side Gas Company Inc.
5010 N. Post Road
Lawrence, IN 46226

Empire Gas Corporation
1700 S. Jefferson St.
Lebanon, MO 65536

Energy Installations Inc.
600 Wilson Road
Columbus, OH 43204

Farmer-Stockman Co-Op Assn.
814 N. Jefferson
Junction City, KS 66441

Foxco Inc.
40 Winn Ave
Winchester, KY 40391

Franger Gas Co. Inc.
27773 C R 10
Elkhart, IN 46514

Future Fuels Inc.
P.O. Box 7000
Bowling Green, KY 42101

GM International Inc.
P.O. Box 1346
Elkhart, IN 46515

H & H Equipment Co. Inc.
16339 Lima Road Box 68
Huntertown, IN 46748

Heetco Inc.
P.O. Box 188
Lewistown, MO 63452

Hoovers of Minneapolis Inc.
215 N. Sheridan
Minneapolis, KS 67467

Industrial Gastruck Inc.
8100 Central Park Ave.
Skokie, IL 60076
(312) 677-8080

L P Propane Service Inc.
20638 Krick Rd
Walton Hills, OH 44146

LP-Gas Equipment Corp
3216 S. Nordic Road
Arlington Heights, IL 60005

LPF Equipment Co. Inc.
16 East Avenue A Box 100
Hutchinson, KS 67501
(316) 663-5086

McTier Supply Company
Box 761
Lake Forest, IL 60045

Millers LP Gas Co.
10122 Smith Calhoun Road
Plain City, OH 43064

NAMOCO
240 E. Lake St.
Addison, IL 60101

Larry Osting & Sons LP Dist.
210 Garrett Street
Princeton, KY 42445

Pearce Bottled Gas Inc.
P.O. Box 66 Hwy 150
Greenville, IN 47124

Peoria County Service Co.
RR1 Box 30A
Edwards, IL 61528

Schultz Gas Service Inc.
2801 E. 175th Street
Lansing, IL 60438

Southwest Gas Equipment Co. Inc.
Box 335 So 83 Hwy
Liberal, KS 67901

Squibb-Taylor Inc.
2840 E. Cook St.
Springfield, IL 62707

Thompsons Gas Inc.
1431 N. Illinois Street
Belleville, IL 62221

Tri State Consolidated Gas Co.
RR4 Hwy 60 East
Paducah, KY 42001

Vandenberg Gas Company
P.O. Box 510A (11150 Chicago Dr)
Holland, MI 49423

Mountain

D & B Propane Systems
6730 S. 400 West
Midvale, UT 84047

Hills Gas & Appliance Co.
708 Daniel St. Box 21058
Billings, MT 59104

Independent Gas Company
13912 Wancrest Drive
Pine Junction, CO 80470

Knudsen Philgas & Oil Co. Inc.
1520 North State
Provo, UT 84601

Propane Carburetion, Inc.
7107 S. 400 West
Midvale, UT 84047

Propane Carburetion Inc.
4250 S. Sante Fe
Englewood, CO

South West

Allen Test Center Inc.
6001 Gibson SE
Albuquerque, NM 87108

American Butane & Propane Gas
1509 Exchange Ave
Oklahoma City, OK 73108

Bms Inc.
3930 Fuqua St.
Houston, TX 77047

Border Butane & Oil Co. Inc.
1025 Wall Road
El Paso, TX 79915

Briscoes LP Gas Service Inc.
Rte 2 Box 560 (448 E. Hwy 152)
Mustang, OK 73064

Carb-Equip Inc.
2640 Northaven #114
Dallas, TX 75229

Enderby Gas Inc.
P.O. Drawer 717
Gainsville, TX 76240

Tom Gorman Company
P.O. Box 15726
Tulsa, OK 74112

Hudson Butane Co. Inc.
1102-10th Street
Bridgeport, TX 76026

Indian Nation Tank Co.
Rt 1
Hugo, OK 74743

Luckys
17500 So Sunnylane
Norman, OK 73069

McCoy Manufacturing Inc.
5705 Azle Ave
Fort Worth, TX 76114

Propane Propulsion Inc.
1150 First National Bank Bldg
Dallas, TX 75202
(214) 225-8111

R & W Supply Inc.
Box 32
Littlefield, TX 79339

Roadrunner Energy Inc.
2621 E. Main
Uvalde, TX 78801

Squibb-Taylor Inc.
P.O. Box 20158
Dallas, TX 75220

Don C. Toomer Co. Inc.
613 South Ave B Box 2319
Freeport, TX 77541

Western

Advanced Propane Services Inc.
12228 SE. Petrovitsky Rd.
Renton, WA 98055

Associated Industrial Products
P.O. Box 14823
Portland, OR 97214

C & J Propane Service
701 Kings Row 29 C
San Jose, CA 95112
(408) 947-7122

Eastern Oregon Propane
Island City Strip Box 943
La Grande, OR 97850

General Propane Corp.
1975 S. Riverside
Bloomington, CA 92316
(714) 877-2642

Mutual Crossman Propane Co. Inc.
210 W. 'B' Street
Wilmington, CA 90744

Mutual Liquid Gas & Equipment
17117 S. Broadway
Gardena, CA 90248

Mutual Liquid Gas & Equipment
1001 E. Edinger
Tustin, CA 92680

Propane Carburetion Inc.
500 Broad
Boise, ID 83706

Propane Conversions Ltd.
2852 Hilltop Dr.
Concord, CA 94520

SCP Gas & Equipment Services
13940 E Dalewood St.
Baldwin Park, CA 91706

Squibb-Taylor Inc.
1880 Deming Way
Sparks, NV 89431

Teeco Products Inc.
16881 Armstrong
Irvine, CA 92714

CARBURETION/EQUIPMENT MANUFACTURERS

Advanced Fuel Components Inc.
904 North Wooded Lane
Marshall, MI 49068

Algas Industries Inc.
P.O. Box 20794
Dallas, TX 75220

Almac Engineering Inc.
20769 Mound Road
Warren, MI 48091

American Carburetion Inc.
823 SW 2nd
Oklahoma City, OK 73109

AmeriGas-LP Gas Division
P.O. Box Z
Paoli, PA 19301

Autogastechniek Holland BV
Glashorst 106
Scherpenzeel 3925 Holland

Ayres Technologies Inc.
224 Datura ST. Suite 1409
West Palm Beach, FL 33401

Beam Products Mfg. Co.
3040 Rosslyn St.
Los Angeles, CA 90065

Beam Products Mfg. Co.
96-10 Linden Blvd.
Ozone Park, NY 11417

Carburetion Equipment Co.
2640 Northaven #114
Dallas, TX 75229

Dix Carburetor Company
10617 S. Pianter Ave.
Santa Fe Springs, CA 90670

Eaton Corp. Controls Div.
191 E. North Ave.
Carol Stream, IL 60187

Empire Gas Corp.
1700 South Jefferson St.
Lebanon, MO 65536

Energy Trends Inc.
180 Research Parkway
Meriden, CT 06450

Facet Enterprises, Inc.
696 Hart St.
Detroit, MI 48204

Fisher Controls Co.
P.O. Box 900
McKinneg, TX 75069

Garretson Equipment Co.
Mt. Pleasant, IA 52641

Gas Supply Inc.
2238 Edgewood Ave. S.
Minneapolis, MN 55426

GCS American Carburetion Inc.
5184 Route 347
Port Jefferson Sta, NY 11776

Holley Carburetor
Colt Industries
11955 East Nine Mile Rd.
Warren, MI 48090

Impco Carburetion Inc.
16916 Gridley Place
Cerritos, CA 90701

Independence Carburetion Systems
215 E. Kansas Ave.
Independence, MO 64050

J & S Carburetor Mfg.
P.O. Box 10391
Dallas, TX 75207

KPF Equipment Co. Inc.
P.O. Box 522
Hutchinson, KS 67501

Landi-Hartog
3208 S. Hutchenson Ave.
Los Angeles, CA 90034

LP-Gas Equipment Corp.
3216 S. Nordic Rd.
Arlington Heights, IL 60005

Marvel-Schebler/Tillotson Division
707 South Side Drive
Decatur, IL 62525

McCoy Manufacturing Inc.
5705 Azle Ave.
Fort Worth, TX 76114

Neptune Measurement Co.
Emerald Rd., P.O. Box 792
Greenwood, SC 29646

Overseas Operations Inc.
P.O. Box 7000-222
Redondo Beach, CA 90277

Petro Systems International
402 Harding Industrial Dr.
Nashville, TN 37211

Petroleum Energy Products Inc.
Route 3 Box 90
Bristow, OK 74010

Propane Equipment Corp.
11 Apple St. Box 836
Tinton Falls, NJ 07724

Ransome Gas Industries Inc.
2050 Farallon Dr.
San Leandro, CA 94577

Rego Company
4201 W. Peterson Ave.
Chicago, IL 60646

Sherwood Selpac
120 Church St.
Lockport, NY 14094

Suburban Manufacturing Co.
Box 399 North Broadway
Dayton, TN 37321

Vialle USA Inc.
7042 Lamson Ave.
Garden Grove, CA 92741

PROPANE FUEL TANK MANUFACTURERS

American Bridge Division of US Steel
P.O. Box 11925
Fresno, CA 92775

American Welding & Tank Co.
P.O. Drawer JJ
Plant City, FL 33566

Bird Automotive
500 Howland St.
Redwood City, CA 94063

Bowen Company
2030 Texas Ave.
El Paso, TX 79901

Brunner Engineering & Mfg., Inc.
P.O. Box 1367
Bedford, IN 47421

Brunner Engineering & Mfg., Inc.
4425 Enterprise St.
Fremont, CA 94538

Cal Tank
P.O. Box 27310
Redding, CA 96001

Dal-Worth
P.O. Box 1018
Grand Prairie, TX 75050

DS Cylinder Inc.
845 Towbin Ave.
Lakewood, NJ 08701

Enderby Gas Inc.
P.O. Drawer 717
Gainsville, TX 76240

Energy International, Inc.
P.O. Box 1346
Elkhart, IN 46515

Fabmet Corp.
One Paul Place
Buffalo, NY 14219

Roy E. Hanson, Jr., Mfg.
1924 Compton
Los Angeles, CA 90011

Indian National Tank Co.
Rt. 1
Hugo, OK 74743

Industrial Manufacturing
P.O. Box 698
Sweetwater, TX 79556

Isometrics, Inc.
P.O. Box 660
Reidsville, NC 27320

Kentwood Quality Service
4630 Danvers
Kentwood, MI 49508

Lee Cylinder Inc.
P.O. Box 287
Cambridge City, IN 47327

Manchester Tank & Equipment Co.
2880 Norton Ave.
Lynwood, CA 90262

Manchester Tank & Equipment Co.
29389 Lexington Park Drive
Elkhart, IN 46514

Manchester Tank & Equipment Co.
2738 Lithonia Industrial Blvd.
Lithonia, GA 30058

Manchester Tank & Equipment Co.
P.O. Box 1338
Lubbock, TX 79408

Master Tank & Welding
P.O. Box 39
Quincy, FL 62301

Melben Products Co., Inc.
100 Production Drive
Harrison, OH 45030

National Butane Gas Co.
P.O. Box 14626
Memphis, TN 38114

Quality Steel
P.O. Box 249
Cleveland, MI 28732

Riley Beaird
P.O. Box 31115
Shreveport, LA 71130

Texel Manufacturing
P.O. Box 1274
Lubbock, TX 79408

Trinity Industries
P.O. Box 10507
Dallas, TX 75207

Wendland Manufacturing
P.O. Box 698
San Angelo, TX 76901

PROPANE PUBLICATIONS

Butane-Propane News
P.O. Box 419
Arcadia, CA 91006

LP-Gas Magazine
1 E. 1st St.
Duluth, MN 55802

FIRE CODE INFORMATION

National Fire Protection Association Inc.
(NFPA #58)
Batterymarch Park
Quincy, MA 02269

Index

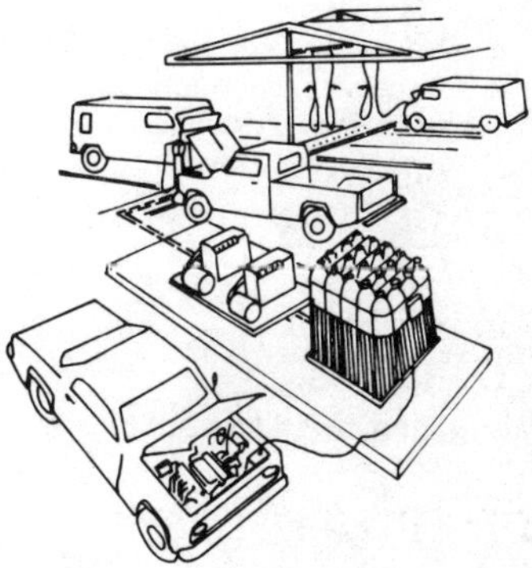